This book presents a multidisciplinary overview of how and why human populations vary so markedly in their skin colour. The biological aspects of the pigment cell and its product melanin are reviewed. Pigmentation in organs other than skin (e.g. eye, ear and brain) are considered as are the common pigmentary disorders. Detailed reflectance data from worldwide surveys of skin colour are also presented. The historical and contemporary background of the phenomenon is explored in sociological terms. Finally, the possible evolutionary forces which shape human pigmentation are assessed. This fascinating account will be of interest to graduate students and researchers of biological anthropology, anatomy, physiology and dermatology, as well as to medical practitioners.

Cambridge Studies in Biological Anthropology 7

Biological perspectives on human pigmentation

Cambridge Studies in Biological Anthropology

Series Editors

G.W. Lasker
Department of Anatomy, Wayne State University,
Detroit, Michigan, USA

C.G.N. Mascie-Taylor
Department of Biological Anthropology,
University of Cambridge

D.F. Roberts
Department of Human Genetics,
University of Newcastle-upon-Tyne

Also in the series

G.W. Lasker *Surnames and Genetic Structure*
C.G.N. Mascie-Taylor and G.W. Lasker (editors) *Biological Aspects of Human Migration*
Barry Bogin *Patterns of Human Growth*
Julius A. Kieser *Human Adult Odontometrics*
J.E. Lindsay Carter and Barbara Honeyman Heath *Somatotyping – Development and Applications*
Roy J. Shephard *Body Composition in Biological Anthropology*

Biological perspectives on human pigmentation

ASHLEY H. ROBINS

Senior Lecturer and Senior Consultant in Clinical Pharmacology,
University of Cape Town Medical School and Groote Schuur Hospital,
Cape Town, South Africa

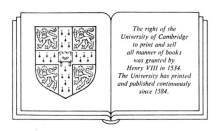

The right of the
University of Cambridge
to print and sell
all manner of books
was granted by
Henry VIII in 1534.
The University has printed
and published continuously
since 1584.

CAMBRIDGE UNIVERSITY PRESS

Cambridge
New York   Port Chester
Melbourne   Sydney

Published by the Press Syndicate of the University of Cambridge
The Pitt Building, Trumpington Street, Cambridge CB2 1RP
40 West 20th Street, New York, NY 10011-4211, USA
10 Stamford Road, Oakleigh, Melbourne 3166, Australia

© Cambridge University Press 1991

First published 1991

Printed in Great Britain at the University Press, Cambridge

British Library cataloguing in publication data
ROBINS, Ashley H.
  Biological perspectives on human pigmentation.
  1. Men. Skin. Pigmentation
  I. Title
  612.7927

Library of Congress cataloguing in publication data available

ISBN 0 521 36514 7 hardback

Contents

To my father and in memory of my late mother

Preface

Benjamin Franklin is alleged to have written: 'But in this world nothing can be said to be certain, except death and taxes'. And one is sorely tempted to add 'the colour problem'. As black people have discovered during this and previous centuries, skin colour is the most decisive and the most abused of all the physical characteristics of humankind. It determines social perceptions, value judgments and interpersonal relationships, and it can wreak havoc on an individual's sense of dignity and self-esteem.

In this book I have endeavoured to analyse the essential nature and functions of human skin colour. I have done this predominantly from a biological standpoint, although I have included a chapter on the psychosocial dimensions of the subject and also one on the possible evolutionary forces which have determined skin colour variations among populations in different geographical regions. Disorders of pigmentation receive special attention, and I have given fairly detailed consideration to the pigmentation that occurs in sites other than the skin and hair.

The field of melanin pigmentation in all its guises is awash with journal articles, monographs and books. I have generally restricted references either to the original authors or to updated reviews, as it would have been unnecessarily cumbersome to cite the multiplicity of contributors to a particular topic. An exception is where the matter under discussion is controversial, or where I expose a personal viewpoint (as I do in assessing the vitamin D hypothesis of skin depigmentation). Here I have felt obliged to furnish fuller documentation for the arguments advanced.

The problem of race and racial labelling has been one of the most taxing for me. The current *Zeitgeist* in physical anthropology is to jettison the idea of race as a biological entity. The reason is that race is arbitrary; racial categories are not clearly circumscribed and certain populations defy classification. Certain physical features, such as skin pigmentation, are not fixed racial characters but rather adaptive traits to suit a particular climate or environment. For example, the juvenile blondness of the otherwise very dark Aborigines of the Western Desert of Australia (which, as Carleton Coon noted, is not due to a Viking invasion!) creates disorder among those committed to racial typologies.

Racial groupings are so enshrined in our ways of thinking that it would have been impracticable in a volume on skin colour to dispense with racial designations. I have therefore not done so, but I must emphasize that this in no way implies my endorsement of the biological validity of racial divisions.

The assignment of names to the so-called races has been a vexatious one, compounded by the political overtones of words such as 'Negro'. In general I have adhered to the traditional anthropological nomenclature, namely, 'Negroid', 'Caucasoid', 'Mongoloid' and 'Amerindian' (while recognizing the limitations of such terms), but where the book deals with historical and socio-psychological issues I freely use the words 'black' and 'white' as they flow better with the less formal style of these sections.

Acknowledgements

I wish to record my gratitude to Groote Schuur Hospital and the Hospital Services division of the Cape Provincial Administration for granting me the six months of special leave that made it possible for me to write this book.

I appreciate the cooperation of all those people who so willingly provided me with photographic material. (Their names have been cited after the legends to the photographs.) Many of these photographs demonstrate rare or unusual clinical conditions and it has been particularly generous of the donors to allow me to reproduce them. My special thanks go to Dr Jennifer Kromberg of the Department of Human Genetics, University of the Witwatersrand, Johannesburg, for her group of photographs relating to albinism and for the very informative discussions that I had with her. I am also indebted to Professor Norma Saxe, Head of the Department of Dermatology, Groote Schuur Hospital, for allowing me to select items from her remarkable slide collection of dermatological disorders. Professor Alan Morris and Professor Phillip Tobias gave me helpful suggestions concerning certain aspects of the book.

I am very grateful to Valerie Myburgh for her meticulous artwork and to Faldiela Martin for excellent secretarial assistance.

Finally, I owe an enormous debt to my wife, Edna, not only for her advice and wisdom but for her unflagging care and encouragement during my periods of writing and rewriting, which were always long, often trying, and sometimes desperate.

1   *Biology of the pigment cell*

Architecture of the skin

An understanding of the biology of skin pigmentation requires some knowledge of the structure of the skin. There is a tendency to regard skin merely as the integument for the otherwise intricate and intriguing machinery of the human body. Yet, in terms of its multiplicity of cellular and fibrous constituents, the skin is not only the largest and most versatile organ of the body but also, with the possible exception of the brain, the most complex.

Dermis

The skin (Fig. 1.1(*a*)) has two major components, the *dermis* and the *epidermis*. The dermis is basically a connective tissue layer comprising collagen, elastic and reticular fibres. It is traversed by a rich network of blood and lymphatic vessels. It also contains structures originally derived from the epidermis – the sweat and sebaceous glands, the hair follicles, and the hairs themselves. Attached to the hair follicles are minute bundles of smooth muscle (*arrector pili*), the contraction of which during cold or fear produces the phenomenon of 'goose-flesh'. The dermis is supplied with sensory nerve endings (mediating the sensations of touch, heat, cold and pain) and with sympathetic nerves which regulate the activity of the sweat gland, arteriole and arrector pili.

Epidermis

The epidermis is a thin layer (about 0.10–0.15 millimetres in thickness) devoid of either blood or nerve supplies. It is composed of two distinct cell populations – epithelial cells or *keratinocytes* (also known as *Malpighian cells*) and pigment cells or *melanocytes*. Keratinocytes comprise about 95 per cent of the epidermis and are arranged in four layers or strata (Fig. 1.1(*b*)) which, moving from within outwards, are:

(a)   *basal layer* (stratum basale or stratum germinativum) – this is a

1

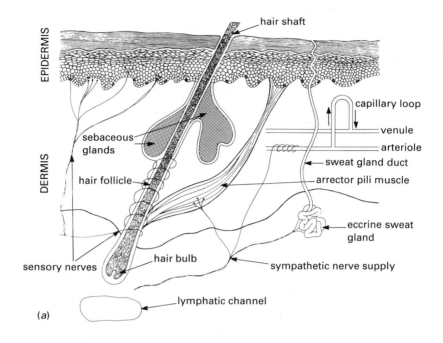

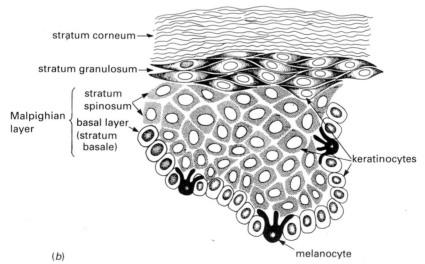

Fig. 1.1(*a*). Structure of skin with particular reference to dermis and its components. Connective tissue fibres such as collagen (abundantly present in dermis) are not shown.

(*b*) More detailed structure of epidermis showing its four layers. The stratum lucidum, present only in thick epidermis (e.g. palm, sole), is not included.

single layer of columnar cells lining the dermal marginal surface. Melanocytes constitute approximately 10 per cent of the
cells in this layer, and melanin is frequently present in the basal
keratinocytes (particularly of black skin).

(b)   *stratum spinosum* – this is several layers thick and comprises
irregular polyhedral cells which become somewhat flattened on
their outer edges. These cells are sometimes called 'prickle
cells', as their surfaces are covered with short spines or projec-
tions which form bridges with adjacent cells.
Some authors use the term *Malpighian layer* to refer jointly to
the basal layer and the stratum spinosum.

(c)   *stratum granulosum* – this consists of several layers of flattened
polyhedral cells whose long axes lie parallel to the skin surface.
Their cytoplasm contains granules of *keratohyalin* (which
apparently contribute to keratin formation), and as these granules increase in size and number so the cell nuclei gradually
degenerate and the cells die.[1]

(d)   *stratum corneum* – this is composed of a varying number of
layers of dead keratinized cells lying closely applied and fused
to one another, except on the outer edge where desquamation
of individual flakes takes place.

Thus, the epidermal keratinocytes undergo characteristic changes as they
are progressively shifted upwards from the basal epidermal layer to the
stratum corneum. The constant shedding of dead horny cells from the
latter is balanced by the formation (through mitosis) of new keratinocytes
in the basal layer. Under normal circumstances, the time required for a
keratinocyte to negotiate all stages from basal layer to stratum corneum is
between 4 and 6 weeks.

Historical aspects of pigment cell biology

Human skin colour, although influenced to a minor extent by pigments
such as carotene, reduced haemoglobin and oxyhaemoglobin, is predominantly based on the pigment *melanin*, a term derived from the
Greek word *melas* (black).

The biological study of melanin pigmentation probably dates back
to Alexis Littre (1658–1726), a French surgeon who noted that the colour
of black skin resided in what was then known as the reticular layer of
the skin (now called the Malpighian layer). A more detailed survey of
racial pigmentation was made by the Leyden anatomist, Albinus, in
his book *Sede et causa coloris Aethiopum et Caeterorum Hominum*

(1737). Albinus described regional differences in skin pigmentation and observed that the colour of the reticular layer corresponded in intensity to skin colour.

The two most important eighteenth-century contributions to the study of human pigmentation came from John Mitchell in Virginia, America, and Le Cat in France. Le Cat produced a comprehensive state-of-the-art review (including evolutionary, anatomical, physiological and clinical aspects) in his monograph entitled *Traité de la couleur de la peau humaine* (1765), while Mitchell in 1745 wrote what is probably the first treatise from the New World on ethnic pigmentation. Mitchell brilliantly applied the principles of Newtonian optics and colour theory to human skin colour, and in several ways he foreshadowed some of our current concepts. The ideas of these four investigators are explored more fully in Wassermann (1974).

With the nineteenth century came the advent of the microscope and early attempts by scientists to examine the microscopic features of pigmented tissue. Although hampered by primitive instruments and the lack of staining techniques, they managed to identify pigment cells in the Malpighian layer and pigment granules in various epidermal sites. In 1826 Laennec made his historical report on melanoma, which he called 'la melanose', although descriptions highly suggestive of melanoma had appeared as early as 1659 (Becker, 1959). Robin is claimed to have been the first to use the term *melanin* when, in 1873, he named the pigment in the pigment cells of animals 'pigment melanique' (Becker, 1959).

The twentieth century witnessed the pioneering histochemistry of Bruno Bloch who, in 1917, placed human skin sections in an aqueous solution of 3,4-dihydroxyphenylalanine (dopa) and demonstrated the appearance of blackened dendritic (branching) cells at the epidermal–dermal junction. He called these cells 'melanoblasts', established the so-called 'dopa oxidase' reaction, and formed the hypothesis that dopa oxidase was a specific intracellular enzyme that catalysed the oxidation of dopa to melanin in humans. Raper (1928) showed that tyrosine was the first compound in the melanin pathway and that it was oxidized by the enzyme *tyrosinase* to dopa, which was then converted to melanin through a series of intermediates.

The current era of pigment-cell biology began after the Second World War when an international community of scientists initiated research in earnest and at a sophisticated level. The First Pigment Cell Biology Conference took place in 1947 and to date there have been 14 such conferences. In July 1987 a new journal, *Pigment Cell Research*, was launched. This reflects the marked popularity of the pigment cell as a focus for investigation, in which respect it must rank next to the red blood cell and the neurone.

There have been a number of dedicated and distinguished researchers in the pigment cell arena, but Thomas B. Fitzpatrick (born 1917) of Harvard University Medical School emerges as the person with the most outstanding and consistent record of achievement over four decades. He and his colleagues have made seminal contributions to melanocyte biology and pathology, and many of the developments and advances reported here have emanated from his laboratories.

Melanocytes
Origin and migration

The cellular basis of melanin formation in mammals was resolved by Billingham (1948), who proved that melanin was synthesized only in the cytoplasm of specific dendritic cells (*melanocytes*) and noted that the dendrites of melanocytes made contact with the surrounding keratinocytes into which they discharged their specialized contents (melanin).

The origin of melanocytes had been the subject of heated controversy for nearly half a century. One standpoint was that melanocytes were indigenous to the epidermis and that they were indeed keratinocytes modified to produce melanin: each keratinocyte had the potential to become a melanocyte. The alternative view was that epidermal melanocytes were immigrant cells and functionally quite different from keratinocytes. The conflict was eventually resolved by the extensive and elegant grafting experiments of Mary Rawles in mice. She established beyond doubt that mammalian melanocytes are not part of the epidermis but originate from the *neural crest* (Rawles, 1953).

In the embryo all brain and nervous tissue ultimately develops from cells of the neural plate. The neural plate gives rise to the neural tube and, after closure of the neural tube, to a band of cells known as the neural crest. The neural crest gives rise to different types of cells, including the dorsal root ganglia of the spinal cord, the adrenal medulla and certain components of peripheral nerve fibres (Schwann cells). But, of importance to this discussion, the neural crest also generates those cells that are destined to differentiate into the melanocyte series. Prospective melanocytes, known as *melanoblasts*, arise from the neural crest in the second month of human embryonic life and migrate from the head region along either side of the spinal cord to the skin. They enter the dermis, epidermis and hair follicles, and differentiate into melanocytes. These cells populate the dermis in increasing numbers between weeks 10 and 12 of development. From 12 to 14 weeks they usually make their first appearance in the epidermis, although epidermal melanocytes have been identified in the embryo at as early as 8–10 weeks (Sagebiel & Odland, 1972).

Initially the melanocytes are located in the superficial zones of the epidermis but, after the sixth month of foetal life (when their numbers have stabilized), they become established at the epidermal–dermal junction. Melanin synthesis occurs from the fourth or fifth foetal month, although the first signs of it have been detected in a 10-week embryo (Sagebiel & Odland, 1972). The dermal melanocytes decrease in number during gestation and have virtually disappeared at birth with the exception of certain sites, notably the lumbosacral area of some individuals, where their presence manifests as the 'Mongolian spot' (see pp. 128–30).

In addition to the skin melanoblasts are disseminated from the neural crest to other sites – most important of these are the uveal tract of the eye (which includes the iris but not the retina), the inner ear, mucous membranes (particularly of the mouth) and leptomeninges (membranes enveloping the brain and spinal cord). The retinal pigment layer of the eye also contains melanocytes, but these derive from the outer layer of the optic cup and not from the neural crest.

Abundance and distribution

As indicated above, the melanocyte is the distinctive and specialized epidermal cell in which tyrosinase-mediated synthesis of melanin takes place. Melanocytes are situated on the basement membrane at the epidermal–dermal junction and they project their dendrites vertically and horizontally within the epidermis. If viewed under the light microscope melanocytes appear as 'clear' cells and, as Bloch showed, it is only if skin is incubated in dopa that they become diffusely darkened. With the combined techniques of splitting dermis from epidermis (by means of the digestive enzyme trypsin) and of dopa incubation, active melanocytes (known as *dopa-positive* melanocytes) in skin sections can be seen and counted microscopically.

Data on the frequency and distribution of adult human melanocytes have shown a symmetrical distribution over the body but a considerable individual and regional variation. Such is the variation between different studies that the absolute values of melanocyte counts have little meaning. Where there has been consensus it is in highlighting the regional differences in melanocyte density – the face and genital areas have a far greater melanocyte concentration than the trunk and thigh (Szabo, 1954; Staricco & Pincus, 1957; Rosdahl & Rorsman, 1983).

A microscopic study of newborn foreskin (obtained during routine circumcisions) showed that even at birth there is pigmentation of the foreskin, which contains a comparable number of melanocytes to that of the adult (Glimcher, Kostick & Szabo, 1973). There is some relationship between adult melanocyte density and cumulative sun exposure: the density in the exposed lateral aspect of the arm is considerably higher than in the adjacent but unexposed medial area (Gilchrest, Blog & Szabo, 1979). However, even in the newborn infant (before there has been any sun exposure) the forehead is already more pigmented than the medial arm area (Walsh, 1964; Post *et al.*, 1976). Thus 'regional variations' in melanocyte numbers appear to be both inherent and sun-induced.

The average population density of melanocytes is between 1000 and 1500 cells per square millimetre. The total population of active melanocytes in the epidermis of an average-sized Caucasoid man would be 1.8×10^9–2.7×10^9 (Rosdahl & Rorsman, 1983): this could be compacted into an estimated tissue mass of 1.1–1.6 cubic centimetres, with a weight of the order of 1 gram. This weight is a miniscule fraction (0.0015 per cent) of body weight and it is highly disproportionate to the profound impact of the melanocyte system on the human situation.

It is an interesting observation that there is generally no correlation between melanocyte numbers and racial pigmentation and that the body-region differences in melanocyte densities pertain to all races. The cheek of a Caucasoid, for example, will contain more melanocytes per square millimetre than the chest or abdomen of a Negroid. Racial skin colour differences do not, therefore, arise from differences in melanocyte numbers but from other factors to be discussed later. An exception to this is that Australian Aborigines and Solomon Islanders (Melanesians) have significantly increased melanocyte numbers in the forearm skin compared with Australian Caucasoids (Mitchell, 1968; Garcia *et al.*, 1977). This difference may partly relate to the much greater duration and intensity of solar exposure in the non-Caucasoid groups.

Functional significance

The melanocyte (Fig. 1.2) has a single and specific function, namely the manufacture of the pigment melanin. The Canadian worker, Pierre Masson (1948), revealed that the melanocyte is not only a melanin-producing factory but also has a glandular role: its product is *secreted* into other cells, a process he called 'cytocrine'.

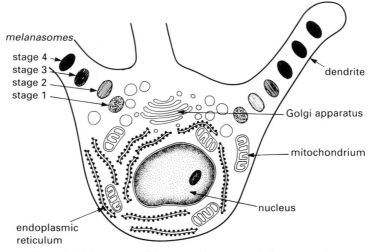

Fig. 1.2. Melanocyte and contents. Melanosome is shown in its developmental stages as it moves peripherally along dendrite. (Based on Jimbow *et al.*, 1976.)

Melanosomes

The basic currency of mammalian pigmentation is a cytoplasmic organelle formerly called a 'melanin granule' but now renamed a *melanosome*. Fitzpatrick *et al.* (1950) were the first to show that the enzyme tyrosinase, which Raper (1928) had identified in plants and insects, occurs in human skin where it converts tyrosine to the brown granular melanin pigment. Birbeck & Barnicot (1959) noted that the melanocyte has a well-developed endoplasmic reticulum (area of protein synthesis) and that it contains melanosomes in all their various stages of maturation. Seiji, Fitzpatrick & Birbeck (1961), by using ultracentrifugation to separate the subcellular components of melanocytes, established that the melanosome itself is the seat of melanin synthesis.

The melanosome begins its life cycle as a spherical, membrane-limited and non-melanized body consisting of two main components – a matrix of structural proteins (arranged in concentric sheets or lamellae) and the enzyme tyrosinase. The precise manner in which the structural proteins and the tyrosinase are fused and organized into a melanosomal unit is still contentious. According to earlier theories, the structural proteins and the tyrosinase (all derived from the endoplasmic reticulum and/or Golgi apparatus) are simultaneously incorporated into the melanosome at its inception, although the tyrosinase remains inactive until melanin synthesis begins. The current concept – which has recently been supported by sophisticated experimental evidence (Akutsu & Jimbow, 1988) – is

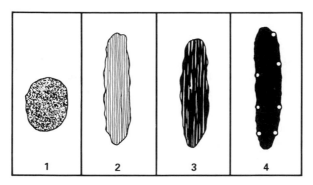

Fig. 1.3 Four stages in melanosome development. Note change in shape from spherical (stage 1) to ellipsoidal and progressive obscuring (stage 3) and ultimate obliteration (stage 4) of internal lamellar structure with increasing melanin deposition. Translucent sacs at outer membrane of stage 4 melanosome represent vesiculo-globular bodies. Although not shown these bodies are also present in stages 1–3. (Based on Quevedo *et al.*, 1987.)

that these two components are synthesized at different sites in the melanocyte. The tyrosinase originates from the Golgi apparatus; the structural proteins of the inner lamellae are assembled in the dilated tubules of the endoplasmic reticulum, the tips of which bud off to become stage 1 melanosomes. Coated vesicles containing tyrosinase are then transferred from the Golgi area to these early melanosomes into which they are taken up and embedded. Melanin formation on the melanosomal lamellae commences only after the tyrosinase has been released from its coated vesicles.

After its formation, the melanosome undergoes a change in shape from spherical to ellipsoidal and a change in appearance from an electron-translucent to an electron-opaque particle. The latter change occurs with the increasing deposition of melanin (melanization) which accompanies the movement of the melanosome from the cell body to the tips of the dendrites (Fig. 1.2). These changes are characterized as four stages by electron microscopy (Quevedo *et al.*, 1987) (Fig. 1.3):

> *stage 1*: the spherical membrane-bound vesicle described above
> *stage 2*: organelle elongated, internal structure shows regular pattern of longitudinal filaments (lamellae) with or without cross-linking
> *stage 3*: organelle still oval in shape, as in stage 2, the internal lamellar structure partially obscured by deposition of electron-dense melanin

stage 4: oval organelle now so electron-dense that internal structure no longer discernible.

Although apparently amorphous and completely opaque these fully melanized melanosomes actually contain minute spherical electron-translucent structures (40 nm diameter) – 'vesiculo-globular' bodies – whose role is unclear. One proposal is that they represent the carriers of tyrosinase (coated vesicles) which, after melanization is completed and the tyrosinase content depleted, degenerate into empty shells within the melanosomes (Jimbow *et al.*, 1976).

Melanocyte–keratinocyte relationship

The melanocytes, which are located on the basal layer of the epidermis, insinuate their dendrites into the surrounding keratinocytes. The association of melanocyte and keratinocyte is far more than just an anatomical juxtaposition – it does, in fact, constitute a functionally-active partnership in which these two cell types are mutually dependent. Fitzpatrick & Breathnach (1963) conceived of the melanocyte and keratinocyte as a structural and functional unit which they termed the *epidermal melanin unit* (Fig. 1.4).

An epidermal melanin unit consists of one epidermal melanocyte in association with the 20–40 keratinocytes to which it donates melanosomes. The number of active epidermal melanin units exhibits a similar regional distribution to that of the melanocytes, but the keratinocyte: melanocyte ratio remains constant.

The process of melanosome transfer from melanocyte to keratinocyte is a crucial one because skin will not appear pigmented unless melanosomes are present within the keratinocytes. Skin and hair colour are perceived by virtue of the melanin content of keratinocytes and not that of melanocytes. Thus any interference with this mechanism of intercellular transfer will lead to hypopigmentation.

The *modus operandi* of this transfer has been intensively studied, and several models proposed. A popular model (and certainly a picturesque one) has been displayed with time-lapse cinematography in epidermal cell cultures (Klaus, 1969; Okazaki *et al.*, 1976). The tip of a melanocyte dendrite (containing melanosomes) becomes enfolded in the recipient keratinocyte. This tip is then nipped off together with its cluster of melanosomes by a process akin to phagocytosis. The melanosome cluster is buried in a cytoplasmic matrix, while the phagocytosed dendrite is gradually decomposed with the eventual release and dispersal of its melanosomes. This scheme has been criticized because there is no

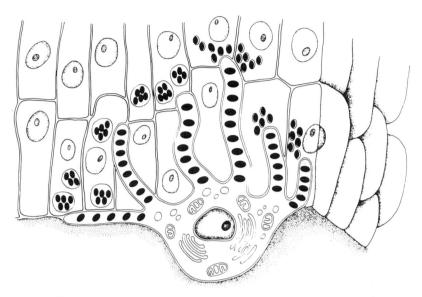

Fig. 1.4 Epidermal melanin unit, showing the structural and functional relationship between melanocyte and surrounding cluster of keratinocytes to which it transfers melanosomes. In Caucasoid and Mongoloid keratinocytes (left) melanosomes are aggregated in groups; for within Negroid keratinocytes (right) melanosomes are singly dispersed (see also Fig. 1.6). (Modified from Fitzpatrick & Breathnach, 1963 and Jimbow *et al.*, 1976.)

certainty that the melanosome cluster is truly intracytoplasmic (Garcia, Flynn & Szabo, 1979).

A second model for melanosome transfer postulates the direct secretion ('injection') of melanosomes by dendrites into the keratinocyte cytoplasm. This is accomplished by intimate cell-to-cell contacts and the creation of cytoplasmic bridges by fusion of keratinocyte and melanocyte membranes (Garcia *et al.*, 1979).

Another model (for which there is little experimental evidence) involves the extrusion of melanosomes from the tips of dendrites into the intercellular space from which they are subsequently removed when engulfed by keratinocytes.

Although the mode of transfer is still debatable, it is clear that keratinocytes are not merely a passive vehicle for the distribution of melanosomes. They appear to regulate several aspects of melanocyte behaviour (e.g. growth, melanization, dendricity), possibly by the release of keratinocyte-derived chemical factors (Gordon, Mansur & Gilchrest, 1989). The influence of keratinocytes has been demonstrated

in a novel study in which cell cultures of Negroid or Caucasoid melanocytes were mixed with cultures of Negroid or Caucasoid keratinocytes.
There was no genetic discrimination between the two cell types: Caucasoid melanocytes donated melanosomes both to Caucasoid and to
Negroid keratinocytes and vice versa (Hirobe, Flynn & Szabo, 1986). In
Caucasoid melanocyte–Negroid keratinocyte co-cultures the Caucasoid
melanocytes became more dendritic and more pigmented than in Caucasoid melanocyte–Caucasoid keratinocyte co-cultures. This experimental
result implies a major feedback mechanism from keratinocytes to
melanocytes.

Distribution of melanosomes within keratinocytes

Once melanosomes are liberated and dispersed within the cytoplasm of
keratinocytes they are arranged in two configurations – either as single
particles (non-aggregated form) or as groups of two or more particles
within a membrane-bound vesicle (aggregated form). The aggregated
forms (also known as 'melanosome complexes') seem to represent
lysosomal structures. The latter contain hydrolytic enzymes, such as acid
phosphatase, which degrade the protein and lipid components of melanosomes but not their melanin. As the basal keratinocytes are pushed
upwards towards the stratum corneum the aggregated melanosomes are
presumed to undergo fragmentation into a fine dust, although this
process may not be so extensive as was previously believed. The singly
dispersed melanosomes may also have lysosomal properties, but they are
more resistant to degradation and arrive at the stratum corneum relatively intact.

The factor which determines whether melanosomes will be consigned
to a solitary or an aggregated state seems to be their size (Toda *et al.*,
1972). Small melanosomes are aggregated and subjected to lysosomal
degradation; large melanosomes are singly dispersed. This aspect of
melanosome distribution will be discussed in more detail in the next
section.

In summary, skin colour is influenced by a spectrum of processes
ranging from the migration of melanoblasts to the disposal of melanin in
the stratum corneum. Basically, the intensity of skin colour is determined
by (a) the total number and size of melanosomes within the epidermal
melanin unit, (b) the rate of melanosome formation and melanization,
and (c) the rate of melanosome transfer to keratinocytes. Other relevant
factors include epidermal thickness, dermal blood supply and the reflective and absorptive properties of skin.

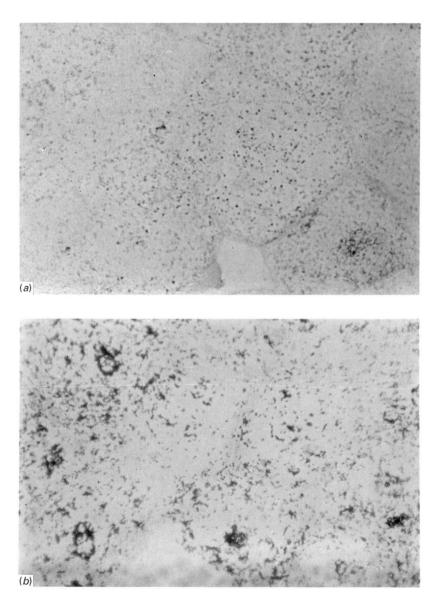

Fig. 1.5 Epidermal stripping from Caucasoid ('white') skin showing small melanosomes sparsely distributed throughout epidermal cells.

(b) Epidermal stripping from 'Cape Coloured' ('brown') skin showing larger and more densely packed melanosomes with some tendency towards central clumping.

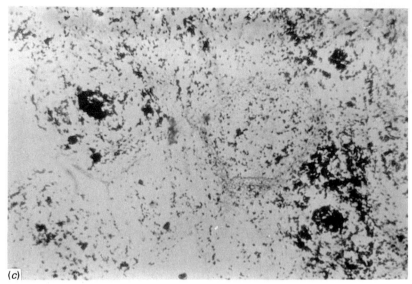

Fig. 1.5(*c*) Epidermal stripping from South African Negroid ('black') skin showing even larger melanosomes and definite evidence of central clumping. (Figs. 1.5(*a*), (*b*), and (*c*) by courtesy of Professor H. P. Wassermann.)

Racial differences in the epidermal melanin unit

As noted above, there are no significant differences in the actual numbers of melanocytes in the skin of different racial groups. The critical factor in the determination of pigmentary variations between and within populations is the melanosome. Microscopic studies of melanosomes in the epidermis of individuals drawn from across the skin colour spectrum have demonstrated quantitative and qualitative differences.

Light microscopy

In Caucasoid skin the melanosomes are small in size and sparsely distributed throughout the epidermal cells (Fig. 1.5(*a*)); in Negroids the melanosomes are larger and more abundant than in Caucasoids, and tend to form clumps centrally within the cells (Goldschmidt & Raymond, 1972; Wassermann, 1974) (Fig. 1.5(*c*)). Mongoloid skin shows almost as many melanosomes as Negroid, but they are noticeably smaller in size, and central clustering occurs only occasionally (Goldschmidt & Raymond, 1972). The epidermis of 'Cape Coloured' individuals contains larger, more numerous and more densely packed melanosomes than Caucasoid skin, with a slight tendency towards central clumping, but on none of these variables do the melanosomes equal those in Negroid skin (Wassermann, 1974) (Fig. 1.5(*b*)).

The most obvious difference seen in light microscopy, therefore, is the abundance of melanosomes in epidermal scrapings from Negroids and their paucity in those from Caucasoids. This observation has some application in forensic science where small skin fragments (or even stratum corneum cells) collected from clothing, scarves or towels may help to identify the skin colour of the individual in cases involving dark-skinned Negroids or fair-skinned Caucasoids (Goldschmidt & Raymond, 1972). There would be far less accuracy or certainty in differentiating between light-skinned Negroids, Mongoloids or deeply tanned Caucasoids.

Electron microscopy
Melanocytes
Generally the melanocytes of very fair-skinned Caucasoids consist of few melanosomes and hardly any in stages 3 and 4. Darker Caucasoids have many more melanosomes, some in stage 4 but most in stages 1, 2 and 3. Mongoloid skin has numerous melanosomes in stages 2, 3 and 4, whereas Negroids have mainly stage 4 melanosomes. Hence, deeper skin colour is associated with a greater number of the more densely melanized melanosomes.

Melanosomes

Size
Table 1.1 lists the sizes of melanosomes in epidermal keratinocytes from Negroid, Mongoloid and Caucasoid skin specimens. There is a tendency for more deeply pigmented skin to contain the larger melanosomes, but it is evident from the data that the size of melanosomes *per se* is neither a distinctive feature of any particular racial group nor necessarily a reflection of the intensity of skin pigmentation. As seen from Table 1.1 moderately pigmented Negroid skin from an unexposed body area has melanosomes similar in size to those from an unexposed, heavily pigmented Caucasoid skin and *smaller* than those from exposed, heavily pigmented Caucasoid skin (Toda *et al.*, 1972). Furthermore, the study of Everett, Nordquist & Wasik (1979) showed that the Caddo Amerindians had larger melanosomes even than dark-skinned American Negroids, although their skin colour is lighter than that of the latter. The keratinocytes of the scalp hair bulb contain very large melanosomes (1.1–1.3 μm $\times$ 0.5–0.7 μm) in all racial groups (Toda *et al.*, 1972).

Table 1.1. *Melanosome distribution and size in skin specimens from different racial groups*

Skin specimen	Melanosome distribution	Melanosome size (μm)	Reference
Negroid skin, moderately to heavily pigmented (unexposed)	Single	1.0–1.3 × 0.5–0.6	Toda *et al.* (1972)
Negroid skin, dark-complexioned (unexposed)	Single	0.69 × 0.28	Olson *et al.* (1973)
Negroid skin, dark[a]	Single and aggregated	Single: 0.43 × 0.17[b] Aggregated: 0.39 × 0.15[b]	Everett *et al.* (1979)
Negroid skin, moderately pigmented (unexposed)	Single and aggregated	0.7–0.9 × 0.3–0.4	Toda *et al.* (1972)
Negroid skin, medium-complexioned (unexposed)	Single	0.53 × 0.27	Olson *et al.* (1973)
Negroid skin, medium[a]	Single and aggregated	Single: 0.35 × 0.16[b] Aggregated: 0.25 × 0.14[b]	Everett *et al.* (1979)
Negroid skin, light-complexioned (unexposed)	Single and aggregated	0.43 × 0.18	Olson *et al.* (1973)
Negroid skin, light[a]	Single and aggregated	Single: 0.31 × 0.14[b] Aggregated: 0.31 × 0.13[b]	Everett *et al.* (1979)
Mongoloid skin (exposed)	Single	1.2 × 0.5	Toda *et al.* (1972)
Mongoloid skin (unexposed)	Aggregated	0.7 × 0.3	Toda *et al.* (1972)
Amerindian skin[a,c]	Single and aggregated	Single: 0.32–0.43 × 0.16–0.18[b,d] Aggregated: 0.33–0.41 × 0.12–0.14[b,d]	Everett *et al.* (1979)
Caucasoid skin, heavily pigmented (exposed)	Single	1.2 × 0.5	Toda *et al.* (1972)
Caucasoid skin, heavily pigmented (unexposed)	Aggregated	0.7 × 0.3	Toda *et al.* (1972)
Caucasoid skin, lightly pigmented (unexposed)	Aggregated	0.6 × 0.3	Toda *et al.* (1972)
Caucasoid skin (unexposed)	Aggregated (predominantly)	0.41 × 0.17	Olson *et al.* (1973)
Caucasoid skin (unexposed)	Single and aggregated	Single: 0.24 × 0.16[b] Aggregated: 0.35 × 0.06[b]	Everett *et al.* (1979)
Caucasoid skin (after ultraviolet exposure)	Single and aggregated	Single: 0.34 × 0.14[b] Aggregated: 0.29 × 0.09[b]	Everett *et al.* (1979)

[a] Presumably unexposed, although not stated in published study.
[b] These figures represent the average of the five *largest* melanosomes.
[c] Skin samples were taken from three Amerindian tribes – the Kiowa, Comanche and Caddo.
[d] These ranges in melanosome sizes signify the variations between the Kiowa, Comanche and Caddo Amerindians. The highest figures in all cases pertain to the Caddo.

Distribution within keratinocytes

There tend to be racial differences in the arrangement of melanosomes in keratinocytes (Mitchell, 1968; Szabo *et al.*, 1969) (Figs. 1.5 and 1.6). The melanosomes of Caucasoids and Mongoloids occur in membrane-limited aggregates (melanosome complexes) which are seen from the basal layer of the epidermis to the stratum granulosum. The only difference between Caucasoid and Mongoloid melanosome complexes is that melanosomes are more tightly packed in Mongoloid complexes. In the Negroid and the Australian Aborigine the picture differs in that not only do the basal keratinocytes contain more abundant and larger melanosomes than those of Caucasoids and Mongoloids but also these melanosomes are mainly dispersed in the single state and tend to form 'protective' caps overlying the nuclei. It is speculated that the singly dispersed Negroid and Austra- loid melanosomes probably impart a more uniform, denser skin colour than do melanosome complexes, and that they thereby offer more effective resistance to solar radiation damage (Szabo *et al.*, 1969). The melanosomes in hair bulb keratinocytes of all racial groups exist in the single form (Toda *et al.*, 1972).

An ultrastructural study of the epidermal pigmentary system in Cauca- soid, Mongoloid and Negroid newborn infants (Rosdahl & Szabo, 1976) showed that in the keratinocytes of the first two groups melanosomes were mostly in complexes, whereas in the Negroid keratinocytes they

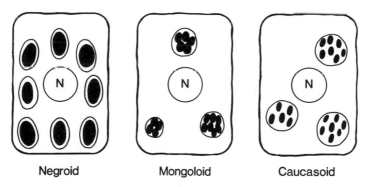

Negroid   **Mongoloid**   **Caucasoid**

Fig. 1.6. Racial differences in arrangement of melanosomes within keratinocytes of unexposed skin. In Negroid keratinocytes melanosomes are large and singly dispersed; in Mongoloid and Caucasoid keratinocytes they are smaller and aggregated into membrane-bound 'melanosome complexes', with ·less ground substance between individual melanosomes in Mongoloid cells. Melanosome distributions are not fixed characteristics of specific racial groups but reflect the predominant pattern. Factors such as UV exposure can alter melanosome configuration. N = nucleus. (Diagram based on Szabo *et al.*, 1969.)

were predominantly single. Negroid neonatal skin contained more numerous and more heavily melanized melanosomes than the Caucasoid, with the Mongoloid skin in an intermediate position. Thus, the principal racial pattern of melanosome number, type and distribution is present at birth and is not determined by ultraviolet-light stimulation.

Determinants of melanosome distribution pattern

Just as the size of melanosomes is not specific to an individual's race so the disposition of melanosomes within keratinocytes is not a fixed characteristic of any particular racial group. Whereas Mongoloids and Caucasoids generally have aggregated melanosomes in their epidermal keratinocytes it is possible, by ultraviolet exposure, to convert this pattern into one with predominantly single melanosomes (Toda *et al.*, 1972). Moreover, as Everett *et al.* (1979) have demonstrated, both single and aggregated melanosomes are observed in all racial groups from light-skinned Caucasoids to dark-skinned Negroids.

The partitioning of melanosomes into either a single or an aggregated state is determined not by race but by melanosome size. The latter factor seems to be the critical one, and a cut-off point was defined by Toda *et al.* (1972). According to their observations melanosomes larger than $0.8 \times 0.3\,\mu m$ were always single, and those smaller than $0.18\,\mu m$ (along the major axis) were always aggregated. Although other workers have not confirmed the diagnostic power of these particular measurements they have highlighted the striking correlation between melanosome size and melanosome distribution (see Table 1.1) – the larger the melanosome (as in Negroids) the more likely it is to be singly dispersed; the smaller the melanosome (as in Caucasoids) the more likely it is to be aggregated in complexes (Konrad & Wolff, 1973; Olson, Gaylor & Everett, 1973; Everett *et al.*, 1979). However, even though the keratinocytes of dark-skinned Negroids were found to contain both single and aggregated melanosomes (Everett *et al.*, 1979) the latter occurred in complexes of only two or three, whereas those of Caucasoids consisted of five or more melanosomes. Thus, Negroids tend to have single melanosomes in their keratinocytes not because they are Negroid but because they tend to have large melanosomes: the converse (aggregated and small melanosomes) applies to fair-skinned Caucasoids. A survey among Solomon Islanders (Melanesians) found that the darker Bougainville Islanders had larger, singly dispersed melanosomes whereas the lighter-skinned Malaita Islanders and Ontong Javanese had smaller melanosomes in melanosome complexes (Garcia *et al.*, 1983).

In general, therefore, there is a relationship between intensity of pigmentation and distribution pattern of melanosomes, in that a dark skin colour tends to be associated with a single disposition and a light colour with complexes. This is not an invariable rule, however, and in Caucasoids with disorders of hyperpigmentation there have been instances of deeply pigmented areas with melanosome complexes and less pigmented ones with single melanosomes (Konrad & Wolff, 1973). Once again, the only correlation that obtained was that between melanosome distribution pattern and melanosome size. This introduces the unresolved dilemma as to whether it is melanosome size or melanosome disposition (single versus aggregated) that is primarily responsible for variations in skin pigmentation (Quevedo *et al.*, 1975).

Hair follicle
The hair follicle is the epidermal tube out of which the hair shaft grows. Hairs are essentially chains of tightly compacted and dead keratinized cells. The hair follicle consists of the hair itself surrounded by an inner and an outer root sheath; its base is expanded in the dermis to form a bulb (Fig. 1.7). It is the cells in the lowest portion (matrix) of the bulb that generate the hair. Hair follicles periodically enter a quiescent phase (telogen) during which they cease hair production; this is followed by an active phase (anagen) when they resume activity. Melanocytes occur in the hair bulb above the tip of the dermal papilla and their activity is closely integrated with that of the matrix cells, so that they become inactive during telogen.

The hair itself, from the outside inwards, comprises the cuticle, the cortex and the medulla. The cuticle represents the skeleton of the hair shaft and its cells are fused with those of the surrounding hair follicle, thereby anchoring the hair securely in position. The cortex forms the bulk of the hair and consists of closely interlocking keratinized cells together with air spaces. The medulla contains large, loosely connected keratinized cells and large air spaces.

In the same way that epidermal melanocytes donate melanosomes to epidermal keratinocytes, melanocytes of the hair matrix transfer melanosomes (during anagen) to the follicular keratinocytes. The melanosomes are dispersed longitudinally in the cells of the cortex and (depending on the depth of pigmentation) also of the medulla. Hair colour is determined by the absorption, reflection and scattering of incident light, and these parameters are influenced by the size, number and distribution of melanosomes. White hair occurs in the absence of melanin and almost all of the incident light is reflected; the shades from blond to black are

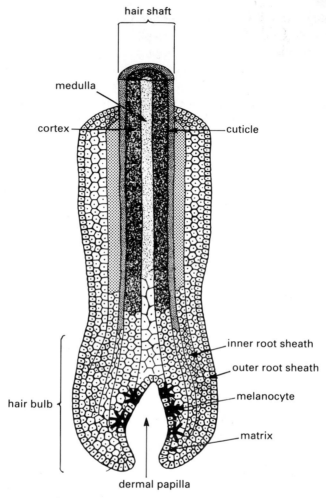

Fig. 1.7. Structure of hair follicle and hair itself, with hair bulb region expanded to show melanocytes and cellular constituents of medulla, cortex and cuticle. As shown, all cells become keratinized to form a solid hair shaft. In pigmented hair, melanosomes are concentrated in the cortex.

associated with increasing melanosome density and increasing absorption of light.

Racial differences in melanosome distribution

A microscopic study by Vernall (1963) examined hairs collected from Negroids, Asiatic Indians, Western Europeans and Chinese. Melanosomes were counted in three regions of each hair: at the outer edge of the cortex (adjacent to the cuticle), at the inner edge of the cortex (bordering on the medulla), and at an intermediate area. Although there was found to be considerable variation within hairs of the same person and between people of the same racial group the greatest variations did occur between subjects of different races. The most consistent finding was that Western Europeans had the lowest densities of melanosomes in all three areas of the cortex. The Chinese had the highest melanosome numbers in the intermediate and inner areas, whereas the Indians had the highest counts in the outer regions. Negroids ranked second in the outer and third in the intermediate and inner areas – in contrast to the epidermal keratinocytes, where they generally surpassed other racial groups in melanosome density. This discrepancy illustrates that the epidermal and follicular melanocyte systems operate independently of each other. A striking example of such disharmony is the infantile blondness observed in the Aborigines of the Western Desert of Australia (Birdsell, 1950), who are among the darkest people in the world. As children these Aborigines have hair as light as the blondest European (Fig. 1.8), and it is only around puberty that their hair colour starts to darken.

Fig. 1.8. Group of Australian Aborigine females showing blondness of hair in youngest member. The blond hair colour usually darkens at puberty. (Photograph: J. B. Birdsell, by courtesy of Professor Phillip V. Tobias.)

Age and the melanocyte system

One of the most distinctive features of human aging is greying of the hair. Age brings about a progressive decline in the number of melanocytes in the hair follicle, with a parallel increase in greying. This decline in melanocytes is not confined to hair but occurs also in the epidermis. Various studies have indicated that functioning (dopa-positive) melanocytes in non-exposed human skin decrease with age by 8–20 per cent of the surviving population each decade (Snell & Bischitz, 1963; Quevedo, Szabo & Virks, 1969; Gilchrest *et al.*, 1979). In sun-exposed or ultraviolet-irradiated skin there are approximately twice as many melanocytes as in unexposed areas, but there is still a comparable decrease in melanocytes with age. The surprising observation is that, unlike hair colour, there is no loss of skin pigmentation with age: on the contrary, the chronically sun-exposed skin of an older person is *more* pigmented than that of a younger subject of similar complexion despite the lower melanocyte density in the former. This paradox has been explained by the greater functional activity (dopa-positivity) in older melanocytes after many years of cumulative sun exposure (Gilchrest *et al.*, 1979).

Genetics of human skin pigmentation

The house mouse has been studied in detail with regard to the genetic control of pigmentation, and it is estimated that at least 150 genes at over 50 loci regulate eye, skin and hair colour. These genes follow a Mendelian pattern of inheritance and produce a wide range of specific effects on such features as the origin and development of melanoblasts, the shape and size of melanosomes, the biosynthesis of tyrosinase and the process of melanosome transfer. In addition, there are many coat colour variants brought about by mutant genes. The 'c' locus in the mouse is believed to control tyrosinase production and thus ultimately the overall depth of pigmentation. Mice homozygous for the recessive 'c' allele are albino, and it is possible that this locus has its counterpart in human tyrosinase-negative oculocutaneous albinism (see p. 142).

Early attempts at elucidating the genetics of human skin colour were frustrated by the lack of objective and reliable measurements of pigmentation. However, since the 1950s the advent of portable reflectance spectrophotometry (see pp. 99–103) has made quantitative studies possible and thereby provided a more precise basis for evaluating the genetic determinants of skin colour variations between and within population groups. These genetic factors have been reviewed by Harrison (1973), Roberts (1977) and Byard (1981).

The *heritability* of skin colour in a population (i.e. the portion of the total phenotypic variability that is genetic) has been estimated by deriving correlation or regression coefficients within members of a family. Such estimates have tended to be inconsistent and problematic, and it is obvious that skin colour is a complex phenomenon which varies in its heritability between different populations (Byard, 1981). However, some indication of heritability is available from analyses of Sikh families (Roberts, 1977) and of Peruvian Mestizo families (Frisancho, Wainwright & Way, 1981) which have yielded estimates of 60–80 per cent and 55 per cent respectively. Post & Rao (1977), using path analysis of skin reflectance measurements from 154 Negroid and 191 Caucasoid American same-sex twin pairs, estimated the total heritability of the skin colour phenotype as 72 per cent, of which race explained 67 per cent and residual genetic factors only 5 per cent. Thus they showed that the only genetic factors which really contributed to skin colour were also the ones that determined race – genetic differences within groups were minor compared with those between groups. It was clear, too, that environmental factors played a substantial role in skin colour determination.

Essentially, the genetics of normal skin colour variation seem best related to a polygenic model with several major genes. Much of the work in this area has compared peoples of Negroid and Caucasoid descent. Stern (1970) took the available information (based on colour top measurements) on the skin colour variations of the Negroid American population and, by constructing theoretical models based on up to 20 gene pairs, sought the one that best fitted the observed data. Using the estimate of a 20 per cent admixture of Caucasoid genes in the Negroid American, he found that models involving three or four additive gene pairs generated a distribution closest to the observed one. Harrison & Owen (1964) studied Caucasoids and Negroids (West Africans), Caucasoid–Negroid hybrids, and backcrosses of these hybrids with either Caucasoids or Negroids. They concluded that three or four gene pairs were responsible for the differences in pigmentation between Negroids and Caucasoids. Subsequent work on the Brazilian Negro (who has an estimated 50 per cent Caucasoid admixture) confirmed that four pairs of genes can account for these differences (Harrison *et al.*, 1967).

The above inferences do not necessarily apply to variations in skin colour involving Caucasoids and other ethnic groups. For example, European and Indian (Asian) differences have been attributed to as many as five gene pairs (Kalla, 1968) or to as few as one or two (Roberts, 1977). The conclusions are conflicting, and indeed the method of constructing models (to determine the number of loci in a hybrid population) has been criticized (Byard & Lees, 1981). There is also uncertainty as to the differential genetic contribution to pigmentation at unexposed and exposed areas. The study of European twins by Clark *et al.* (1981) found that tanning ability was controlled to a lesser degree by genetic factors than was natural pigmentation; however, within Indian (Dehli) families the greater genetic determination occurred at sun-exposed areas (Kalla, 1972; Banerjee, 1984). It appears that clarity on the genetics of human pigmentation will only emerge with the use of more sophisticated methodological approaches (Byard, 1981).

The genetics of human hair and eye colour are complex and various types of inheritance have been proposed. There is no agreement, and the reason for this probably relates to the complexity of these characteristics, their dependence on polygenic mechanisms and the effect of environment. The latter may account for 20 per cent or more of the total variance in eye and hair colour (Bräuer & Chopra, 1980), a surprisingly large contribution for characteristics which are usually presumed to be entirely genetically determined. It does appear that the inheritance of darker hair and eye colour tends to be dominant over lighter coloration. There is sufficient evidence to show that red hair is under genetic control, but the hereditary mechanisms are still obscure. An analysis of families with red-haired members supports the hypothesis that red hair pigment is dominant to blond but hypostatic to brown and black (Rife, 1967).

Note

1 In thick epidermis (e.g. palms and soles) there is an additional layer between the stratum granulosum and the stratum corneum. Known as the *stratum lucidum*, it is translucent, homogeneous and composed of a thin layer of markedly flattened cells, usually devoid of nuclei.

2   *The biochemical and hormonal control of pigmentation*

Character of melanin

Melanins are among the most widespread natural pigments, being present in all living organisms including plants, fungi and bacteria. Plant melanins have a different biochemical derivation from animal melanins. The latter originate from the amino acid tyrosine, and they are characterized by a brown-black colour, a high molecular weight and a polymeric structure. One of the most problematic features of melanin to the scientist is its insolubility in almost all solvents. Because it is stubbornly resistant to chemical treatment melanin is difficult to purify and analyse. Both Littre and Albinus, in the eighteenth century, were amazed at their failure to extract pigment from black skin after subjecting it to prolonged immersion in water and alcohol. Even after two and a half centuries of technology there are still no general methods to solubilize *natural* melanin under physiological conditions. Hence the Harvard biologist Carroll Williams was moved to describe melanin as 'a pigment of the imagination'!

The intractability of melanin, on the other hand, may prove of value to palaeontologists (Daniels, Post & Johnson, 1972). Melanin has been found in a 150-million-year-old ichthyosaur, in extinct mammoth skin and in mummy skin. The cephalopod molluscs possess an 'ink gland', the secretion of which (i.e. melanin) was used by the ancients as a dye (sepia). The squid stores its melanin in a reservoir and, in time of danger, squirts the pigment from a siphon so as to create a smokescreen to divert its enemy. Cephalopod melanin is invulnerable to decay and early in the nineteenth century, in the south of England, a 150-million-year-old squid was discovered whose own ink was used to make drawings of its remains (Fox & Vevers, 1960). Perhaps a specific search for melanin in fossil deposits may provide some clue as to the skin colour of early hominids.

Biochemistry

Melanogenesis

Tyrosinase is the enzyme responsible for the formation of melanin within the melanosomes, a process which will be referred to as *melanogenesis*.

There are two main types of melanin important in human biology – *eumelanin* and *phaeomelanin*.[1] Eumelanin is the black-brown compound that has been discussed thus far: it is found in the skin, hair and all the other melanocyte-bearing tissues. Phaeomelanin is the yellow-to-reddish-brown pigment which has been identified in mammalian hair (including human red hair)[2] and in the feathers of chickens. Phaeomelanin and eumelanin differ in certain chemical and physical properties; for example, only phaeomelanin is soluble in dilute alkali.

Eumelanin

The classical scheme for the conversion of tyrosine to eumelanin has been called the *Raper–Mason pathway*. Tyrosine, in the presence of tyrosinase and oxygen, is oxidized to dihydroxyphenylalanine (dopa) which, in turn, is oxidized (also by tyrosinase) to dopaquinone. Dopaquinone then undergoes a series of spontaneous and non-enzymatic reactions and rearrangements until melanin is produced. The sequence is as follows:

tyrosine $\xrightarrow[\text{oxygen}]{\text{tyrosinase}}$ dopa $\xrightarrow[\text{oxygen}]{\text{tyrosinase}}$ dopaquinone $\longrightarrow$

leucodopachrome $\longrightarrow$ dopachrome $\longrightarrow$ 5,6-dihydroxyindole $\longrightarrow$

indole-5,6-quinone $\longrightarrow$ melanin

It was originally assumed that melanin was a polymeric structure consisting of recurring and inter-linked molecules of indole-5,6-quinone. However, highly complex research done in the 1960s by workers including Swan, Nicolaus, Piatelli and Hempel discredited the notion that melanin was simply a homopolymer of regular indole-5,6-quinone subunits (Mason, 1967). It now seems certain that melanin is a heteropolymer consisting of different subunits. These subunits are the intermediate compounds in the tyrosine-melanin pathway (from dopaquinone onwards) which randomly co-polymerize to form the highly irregular three-dimensional configuration that is eumelanin (Fig. 2.1). Melanin is bound to protein within the melanosomes to become *melanoprotein*, a binding that reinforces the nightmare of chemical analysis.

Phaeomelanin

The biosynthesis of phaeomelanin follows the same route as that of eumelanogenesis up to dopaquinone. Dopaquinone then combines with the sulphydryl-containing amino acid cysteine to form 5-*S*-cysteinyldopa.

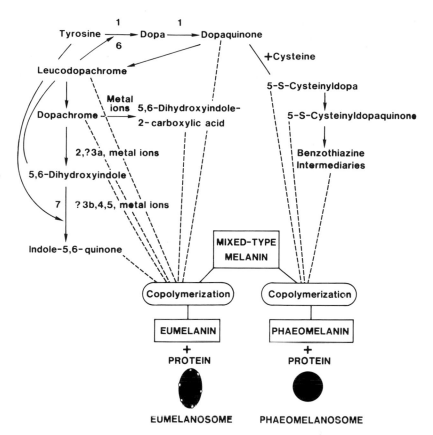

Fig. 2.1 Revised Raper–Mason pathway of melanogenesis. Pathway for biosynthesis of phaeomelanin also shown. Note ellipsoidal shape of eumelanosome with vesiculo-globular bodies, and spherical shape of phaeomelanosome with absence of vesiculo-globular bodies. 1 = tyrosinase (T_4); 2 = dopachrome conversion factor (tyrosinase T_4); 3a = dopachrome oxidoreductase; 3b = dopachrome oxidoreductase (inhibiting reaction in absence of tyrosinase); 4 = indole conversion factor (tyrosinase T_4); 5 = indole blocking factor (tyrosinase T_1 and T_3); 6 = inhibition of reaction by 5,6-dihydroxyindole; 7 = inhibition of reaction by tyrosine.

The latter is oxidized via intermediates to phaeomelanin (Fig. 2.1), which also binds to protein in phaeomelanosomes.

In many mammals the hair is not uniformly coloured. The stripes of the tiger and the spots of the leopard are examples. Sometimes individual hairs are themselves bi-coloured (as in the 'agouti type' hair of many rodents, rabbits and hares): the end portion of the hair contains phaeomelanin and the rest eumelanin. In these cases hair-bulb melanocytes are capable of switching from eumelanogenesis to phaeomelanogenesis, a switch that is genetically determined and is related to the content of cysteine, and probably also of glutathione (another sulphydryl compound), in the melanocyte.

There are ultrastructural differences between eumelanosomes and phaeomelanosomes (Jimbow *et al.*, 1979). The two types are identical at stage 1 (spherical membrane-bound vesicles), but whereas eumelanosomes subsequently become ellipsoidal in shape and show an internal lamellar structure (Fig. 1.3) phaeomelanosomes remain spherical and do not develop lamellae. In their fully melanized form (Fig. 2.1), phaeomelanosomes (unlike eumelanosomes) do not exhibit electron-translucent vesiculo-globular bodies (see p. 10).

Recently another group of melanins (*mixed-type melanin*) has been proposed with properties in common with eumelanin and phaeomelanin. The mixed-type melanin is formed by co-polymerization of the oxidation products of both dopa and cysteinyldopa (Fig. 2.1).

A study of red-headed human subjects (Jimbow *et al.*, 1983) found that some of them possessed only spherical phaeomelanosomes in their hair follicles and others had a mixture of melanosomes, i.e. their follicles contained two different populations of melanocytes – one producing phaeomelanosomes and the other eumelanosome-like granules.[2]

Tyrosinase

This is a copper-containing enzyme believed to play an essential part in melanogenesis. As shown in the Raper–Mason pathway, a curious property of tyrosinase is that it has a dual action. It first catalyses the crucial oxidation of tyrosine to dopa (acting as a tyrosine hydroxylase) and then converts dopa to dopaquinone (acting as a dopa oxidase). Okun *et al.* (1973) caused a stir when they dismissed the role of the so-called 'tyrosinase' as a tyrosine-hydroxylase enzyme and accepted only its dopa-oxidase capacity. They claimed that it was peroxidase, an enzyme also present in the melanocyte, that accomplished the critical tyrosine-to-dopa reaction. Considerable debate ensued, with the weight of scientific evidence generally confirming the original status of tyrosinase as the key enzyme in melanogenesis. A telling argument was that, whether the action of peroxidase was inhibited or stimulated, there was no effect on the hydroxylation of tyrosine to dopa (Pomerantz & Ances, 1975).

Mammalian tyrosinase exists in at least three different molecular forms (isozymes). The present view is that tyrosinase (T_3) is synthesized on ribosomes and transported through the smooth endoplasmic reticulum to the Golgi apparatus where the enzyme is glycosylated to become T_1. The latter is then packaged into coated vesicles and transported to the early-stage melanosomes where, after being membrane-bound, it becomes active (as T_4) in promoting melanogenesis (Hearing, Korner & Pawelek, 1982; Hearing & Jiménez, 1987). The isozymes T_1 and T_3 are regarded as precursors of T_4 and are thought to impede melanogenesis. Hearing *et al.* (1982) have established that the tyrosinase isozymes are involved at the following stages in the Raper–Mason scheme (Fig. 2.1) as regulators of melanogenesis:

(i)   *dopachrome conversion factor* which accelerates the conversion of dopachrome into 5,6-dihydroxyindole: this is associated with the melanosomal tyrosinase isozyme T_4;

(ii)   *indole conversion factor* which accelerates the conversion of 5,6-dihydroxyindole into melanin: this is also associated with T_4;

(iii)   *indole blocking factor* which retards the formation of melanin from 5,6-dihydroxyindole: this is associated with the isozymes T_1 and T_3.

Thus the tyrosinase species (T_1, T_3, T_4) is more versatile than previously imagined, and from catalysing the two original reactions in the Raper–Mason pathway its role has now extended and it seems to be involved in five stages.

In 1984 Barber *et al.* reported on a new enzyme which they called *dopachrome oxidoreductase*. This brought about the conversion of dopachrome to 5,6-dihydroxyindole and, in the absence of tyrosinase, it also blocked the further transformation of 5,6-dihydroxyindole to melanin. Whether dopachrome oxidoreductase corresponds to the dopachrome conversion factor and the indole blocking factor respectively of Hearing *et al.* (1982) is not yet clear. There are even doubts about the actual existence of dopachrome oxidoreductase (Prota, 1988). However, these more recent discoveries refute the pre-1980 concepts which held that all of the steps in the Raper–Mason pathway beyond dopaquinone were non-enzymatic and occurred by spontaneous auto-oxidation.

Korner & Pawelek (1982) showed that 5,6-dihydroxyindole inhibited the conversion of tyrosine to dopa, and that tyrosine inhibited the conversion of 5,6-dihydroxyindole to melanin. They formed the hypothesis that these inhibitory mechanisms provided feedback regulation of mammalian melanogenesis and thereby lessened the potentially cytotoxic effects of melanin precursors (such as the generation of free radicals). A recent proposal is that metal ions (copper, zinc, nickel, iron, etc.), which are highly concentrated in pigmented tissues, are capable of

rearranging dopachrome to 5,6-dihydroxyindole and thereafter of catalysing the oxidative polymerization of the latter to melanin pigment
(Prota, 1988).

Inhibitors of tyrosinase

Rothman, Krysa & Smiljanic (1946) reported the presence in human
epidermis of a sulphydryl compound which inhibited plant tyrosinase.
They postulated that oxidation of this compound, by stimuli such as
ultraviolet light or inflammation, would release the tyrosinase system
from inhibition and cause hyperpigmentation.

Sulphydryl compounds may therefore act as regulators within the
pigmentary process, in that by creating strong bonds with the copper
required for tyrosinase activity they inactivate the enzyme. The more
important mechanism within the melanocyte is probably that sulphydryl
compounds such as glutathione react with the intermediates of the
tyrosine–melanin pathway to form products that do not polymerize (Seiji
et al., 1969).

Skin colour and tyrosinase

Glutathione (a sulphydryl compound) is present in tissue, 90 per cent of it
being in the reduced state. This high level of reduced glutathione is
achieved by the enzyme, *glutathione reductase*, which converts oxidized
to reduced glutathione. Halprin & Ohkawara (1966) claimed that glutathione was the epidermal substance inhibitory to tyrosinase, and
demonstrated that Negroid skin contained less reduced glutathione and
less glutathione reductase than Caucasoid skin (possibly due to genetic
differences in the reductase enzyme). This was the first demonstrable
biochemical difference between skins of different colours.

Confirmation came from data on the tortoiseshell guinea-pig – an
excellent model for a comparative study of pigmentation differences
because the animal has a coat of three colours. It was found that the
lowest levels of reduced glutathione and glutathione reductase were
associated with the black (eumelanin) areas, whereas the highest values
occurred in the yellow or red (phaeomelanin) skin patches (Benedetto *et
al.*, 1981, 1982). When two mutant strains of mice (one black and one
yellow) were compared the observed differences in glutathione and
glutathione reductase showed a similar trend, with the black skin having
significantly lower levels of both compounds.

Another enzyme, *thioredoxin reductase*, has been implicated in the
modulation of melanogenesis. It produces reduced thioredoxin, which is
a direct inhibitor of tyrosinase. The activity of epidermal thioredoxin

reductase has been found to increase with increasing skin pigmentation, and it has been postulated that the enzyme acts to limit the free radicals in the skin generated by melanogenesis (Schallreuter, Hordinsky & Wood, 1987).

Tyrosinase activity has been assayed in Negroid and Caucasoid foreskin samples (obtained from neonates after circumcision) with the finding that Negroid skin had two to three times as much tyrosinase activity as the Caucasoid specimens (Pomerantz & Ances, 1975; Iwata *et al.*, 1990). The correlation between tyrosinase activity and skin melanin content may be due not only to different amounts of enzyme present in Negroid and Caucasoid melanocytes but also possibly to differences in the catalytic activities of the tyrosinase (Iwata *et al.*, 1990).

Investigators have sought to identify metabolic differences between subjects of varying skin colour. Negroids have been found to have a greater urinary excretion of 6-hydroxy-5-methoxyindole-2-carboxylic acid (a metabolite in the eumelanin pathway) than fair-skinned Scandinavians, although there were no differences in excretion levels of 5-*S*-cysteinyldopa (a metabolite in the phaeomelanin pathway) (Wirestrand *et al.*, 1985). Similarly, in a study of Asians and Europeans (including patients with hypopigmentation disorders) it emerged that 5-hydroxy-6-methoxy-2-carboxylic acid levels correlated well with degree of skin pigmentation, whereas levels of 5-*S*-cysteinyldopa did not (Westerhof *et al.*, 1987). It therefore appears that cysteinyldopa excretion is unrelated to skin colour, possibly because some of it may be formed outside the melanocyte. The eumelanin metabolites mentioned above, however, are exclusive to the melanocyte and probably represent the best markers of melanin biosynthesis.

Hormones and skin pigmentation
The Frog

This animal has provided an excellent experimental model to biologists seeking a hormonal basis for skin pigmentation. Unlike mammals, cold-blooded vertebrates such as frogs have a system of melanin-containing cells within the dermis called *melanophores*. Melanophores are capable of producing dramatic colour changes in the animal by means of the rapid and reversible translocation of melanosomes. This flexibility enables the frog to adapt to changes in background coloration. When it is adapted to a white background the melanophore is in a contracted state, with melanosomes aggregated in the central cell area around the nucleus; in the dark-adapted state the melanophore is expanded, with melanosomes dispersed throughout the cell and peripherally in the dendrites.

Melanocyte-stimulating hormone (MSH)

Allen (1916) and Smith (1916) independently reported that removal of the pituitary glands from frogs and tadpoles resulted in skin lightening. It was subsequently shown that frogs immersed in pituitary extracts became darker, because of melanosomal dispersion throughout the stimulated melanophores. Lerner, Shizume & Bunding (1954) demonstrated that the intramuscular injection of an extract of pig pituitary gland into human subjects caused darkening of the skin and naevi. They proposed the term *melanocyte-stimulating hormone* (MSH) for this active pituitary principle. MSH was isolated from the pig pituitary by Lerner & Lee (1955), its structure was elucidated in 1957, and it was synthesized in 1960.

Two forms of MSH were identified in the pig pituitary – *alpha-MSH* and *beta-MSH* – both being polypeptides, but with differing lengths of amino acid chain. There is a close relationship between MSH and another pituitary hormone, *adrenocorticotrophic hormone* (ACTH). ACTH can cause dispersion of melanin granules in frog melanophores, although it has only about 1 and 2 per cent of the potency of alpha-MSH and beta-MSH respectively. Its biological significance in melanin pigmentation is obscure, but in those pathological states where its output is excessive (e.g. Addison's disease), hyperpigmentation is a notable feature. ACTH, alpha-MSH and beta-MSH all include a common sequence of seven amino acids – this 'heptapeptide core' presumably endows the hormones with melanocyte-stimulating activity.

Mode of action of MSH on melanocytes

The administration of MSH induces darkening of mammalian skin and also stimulates the follicular melanocytes with deepening of hair colour (Friedmann & Thody, 1986). Although microscopic studies have revealed that compared with untreated cells treated melanocytes become more dendritic and contain more melanosomes (as do treated keratinocytes), the primary response to MSH is an increase in tyrosinase activity with resultant stimulation of melanogenesis. MSH appears to act by binding to a specific membrane-bound receptor on the melanocyte. This interaction leads to activation of adenyl cyclase, which in turn brings about an increase in the formation of intracellular cyclic AMP (the 'second messenger'). It is this cyclic AMP that relays the effect of MSH and produces an increase in tyrosinase activity (Pawelek, 1976). Whether, in a more general sense, cyclic AMP is an essential or primary regulator of human melanogenesis is unclear, especially in the light of recent work with cultured human melanocytes that suggests a role for protein kinase C as a stimulus for melanin biosynthesis (Gordon & Gilchrest, 1989).

Is there a human MSH?

Alpha-MSH, an important hormone in certain mammals like the rat, is present in human pituitary and human plasma in very small amounts. It had therefore been believed that beta-MSH and ACTH were the pigmentary hormones in the human and, because ACTH was considerably weaker in melanocyte-stimulating activity, beta-MSH had been accepted as the principal pigmentary hormone. However, this position was challenged during the 1970s when complex immunological research was undertaken into the nature of beta-MSH and other pituitary polypeptides.

This research showed that it was not possible to demonstrate the presence of beta-MSH in adult human pituitaries, and that beta-MSH did not in fact exist as such *in vivo* (Brown & Doe, 1978). The human pituitary (anterior lobe) stores two types of molecules which are derived from a common precursor, *pro-opiomelanocortin*. These molecules are ACTH and beta-lipotropic hormone (beta-LPH). The 'beta-MSH immunoreactivity' previously measured in human pituitary and plasma was actually due to beta-LPH, and the so-called 'beta-MSH' was merely a fragment of beta-LPH that was split off from the parent molecule by the proteolytic enzymes used during the extraction procedure. 'Human beta-MSH' was thus simply an artefact, and the term is now relegated to the archive of scientific error! But despite this one still has to explain the unequivocal existence of alpha-MSH and beta-MSH in other mammalian species. These polypeptides are produced in the *intermediate* lobe of the pituitary gland, where the necessary proteolytic enzymes exist to cleave ACTH into alpha-MSH and beta-LPH into beta-MSH. The human pituitary lacks a distinct intermediate lobe and is thus relatively incapable of degrading ACTH and beta-LPH into their respective MSH peptides. The human foetal pituitary, however, has an intermediate lobe (it subsequently involutes) which can produce these peptides (Friedmann & Thody, 1986). It appears that alpha-MSH and beta-MSH may play a part in activating the pigmentary system in the mammalian embryo.

In 1980 a third MSH (gamma-MSH) was purified from human pituitary tissue (Benjannet *et al.*, 1980). This has minimal melanophore-stimulating activity compared with alpha-MSH and like the two other MSH peptides it originates from the pro-opiomelanocortin prohormone (Friedmann & Thody, 1986).

The role of the pituitary pigmentary hormones in the control of *normal* human pigmentation remains an enigma. Albinos have no deficiency of these hormones. Negroid panhypopituitary dwarfs do not lose their pigmentation, nor does surgical removal of the pituitary gland in the Negroid reduce skin colour. ACTH and beta-LPH can cause hyperpigmentation only if present in abnormally high concentrations: under physiological conditions they do not seem to regulate skin pigmentation.

Although the MSH peptides may have lost their pigmentary function during mammalian evolution, recent research has uncovered a variety of other properties such as effects on the immune, cardiovascular and central nervous systems (Eberle, 1988). In its action on brain functions, MSH injection heightens arousal and improves performance in tests of visual retention and reaction time. It seems that the primary central action of MSH (specifically alpha-MSH) is to enhance attention, with secondary effects on memory, learning and behaviour (La Hoste *et al.*, 1980). This mediation in the attentional processes of the cortex has adapted MSH to perform an extrapigmentary role within the sphere of behaviour.

Melatonin

What MSH is to the darkening of frog skin, so *melatonin* is to its lightening. McCord & Allen (1917) noted a remarkably swift, but temporary, blanching of the skin of frogs and tadpoles after feeding them minced mammalian pineal tissue. Lerner *et al.* (1958) reported having isolated melatonin from bovine pineal glands.[3]

Biosynthesis

Melatonin (5-methoxy-*N*-acetyltryptamine) was so named because it causes paling of frog skin by aggregating the melanosomes and so producing a contracted melanophore. It is manufactured in the pineal gland from the amino acid, tryptophan, in a biochemical pathway in which serotonin (5-hydroxytryptamine) is one of the intermediates. The enzyme (hydroxyindole-*O*-methyltransferase) which catalyses the final step to melatonin is found only in the pineal gland of mammals. The activity of this enzyme is regulated by the input of light to the eye. There is an elaborate neural tract which relays light impulses from the photoreceptors in the retina via the hypothalamus and the superior cervical ganglion to the pineal gland. Light decreases the output of noradrenaline (a neurotransmitter) from the sympathetic nerves supplying the pineal and, in so doing, it inhibits the enzyme and reduces melatonin biosynthesis. Darkness, on the other hand, stimulates the enzyme and thereby increases the melatonin level. This light-related cycle means that, in the human, peak melatonin levels are achieved at night during sleep and the lowest levels occur during the day.

Effects on melanogenesis

Melatonin is more potent than any other compound in the lightening of frog skin. It reverses the melanosome-dispersing effect of MSH, and causes melanosomes to stream back to the centres of melanophores making most of the skin transparent. In several respects melatonin appears to be the physiological antagonist of MSH, and these two hormones possibly interact in effecting the adaptive changes of amphibians to background coloration and illumination. Certainly it is in amphibians that the action of melatonin is most conspicuous.

In mammals the effects of melatonin are inconsistent. Daily injections of melatonin into guinea-pigs for one month produced no gross or microscopic effect on either their skin (Snell, 1965) or hair (Clive & Snell, 1969). When five human subjects (Caucasoid) with abnormal skin hyperpigmentation were treated with 1 gram of melatonin daily for 25–30 days only one patient showed any skin lightening: in the others the skin colour remained unchanged (Nordlund & Lerner, 1977). On the other hand, Rickards (1965) injected melatonin into dogs with canine melanosis and noted a lightening of the melanotic areas within 48 hours with desquamation of dark scales, indicating that melanin deposits were being extruded. In contrast, the function of melatonin is more firmly established in those mammals which undergo seasonal changes in their coat colour, changes that are predominantly governed by the length and intensity of daylight. In several of these species subcutaneous melatonin implants inhibit the pigmentation of new hair which grows out following hair plucking or during moulting (Weatherhead, 1982).

Mode of action

The mechanism of action of melatonin is far from understood, but it has well-documented inhibitory actions on reproductive functions. As with MSH a specific receptor has been postulated for melatonin but, unlike MSH, this is not linked to cyclic AMP. Another compound (cyclic GMP) seems to be involved as the 'second messenger'; it apparently acts by inhibiting the later stages of melanin biosynthesis without affecting tyrosinase itself (Weatherhead, 1982).

Role

There is little evidence that melatonin influences human pigmentation. As it is a hormone that is so delicately attuned to the state of environmental lighting its most appropriate niche would be in those animals that rely on seasonal changes in coat colour for adaptive purposes (e.g. concealment or temperature regulation).

The seasonal transformations of the weasel illustrate the complex inter-relationships between light, hair colour, hormones and reproduction. The hypothesis of Quevedo *et al.* (1975) is that decreasing photoperiod in autumn and early winter generates increased melatonin secretion which, in turn, has inhibitory effects on the pigmentary and reproductive hormones of the pituitary. There follows a moult, with regrowth of white hair, and the animal becomes reproductively dormant. With increasing photoperiod during late winter and spring melatonin secretion decreases and these changes are reversed: the spring moult brings regrowth of pigmented hair and reproductive activity resumes.

The daily cyclical pattern in the secretion of melatonin may be related to to its possible function as a 'biological clock' which orchestrates a variety of physiological processes. In contrast with the cerebral-alerting effects of MSH, melatonin has sleep-inducing properties (Nordlund & Lerner, 1977; Lieberman *et al.*, 1984). The ability of melatonin to promote sleep suggests that the hormone might mediate the synchronization of human circadian rhythms to the light–dark cycle. This proposal has been tested by using melatonin in placebo-controlled, double-blind trials on subjects flying across numerous time zones, namely, from San Francisco to London (Arendt, Aldhous & Marks, 1986) and from Auckland (New Zealand) to London and back (Petrie *et al.*, 1989). In these trials melatonin significantly alleviated the jet lag and tiredness occasioned by such flights.

There is a cyclical disorder of mood (*seasonal affective disorder*) characterized by recurrent episodes of autumn/winter depression alternating with periods of spring/summer elation. Affected patients improve, during their depressed winter phases, after exposure to bright artificial light of a far greater intensity (e.g. 2500 lux) than ordinary domestic or occupational lighting.[4] The hypothesis is that the antidepressant effects of bright light in seasonal affective disorder are mediated by the suppression of melatonin secretion which is itself induced by the light.

However, recent evidence from several different experimental approaches has failed to support a decisive role for melatonin (either in the pathogenesis of the disorder or in the antidepressant response to light therapy), although an indirect involvement is not excluded (Rosenthal *et al.*, 1988).

Sex hormones

Oestrogen and progesterone

In guinea-pigs surgical removal of the ovaries (ovariectomy) has been shown to cause shrinkage of melanocytes and reduction of their melanin content, with the area around the nipple (areola) being particularly affected by the loss of melanocytes (Bischitz & Snell, 1960). When these same animals were treated with small doses of oestrogen, skin melanin (especially in the areola) increased. Larger doses had a more pronounced effect on melanogenesis than smaller doses, and also increased the melanocyte count in the anterior abdominal wall (Snell & Bischitz, 1960). In women who had undergone ovariectomy oestrogen administration was found to have increased the skin pigmentation, although this was attributed largely to elevated skin oxyhaemoglobin resulting from a greater cutaneous blood supply (Edwards & Duntley, 1949).

Progesterone expands frog melanophores, but its influence in the mammal is less clear. Small doses caused slight stimulation of melanogenesis in the ovariectomized guinea-pig, whereas larger doses had the opposite effect (Snell & Bischitz, 1960). When given together, oestrogen and small doses of progesterone had a greater stimulatory effect than either given alone.

Androgens

The action of androgens on melanogenesis is even more obscure than that of oestrogen and progesterone (Thody & Smith, 1977). Testosterone has enhanced pigmentation in women, in castrated and hypogonadal men, and in the scrotal skin of monkeys and rats, but there are species differences – guinea-pig skin, for example, darkens after castration and lightens after treatment with testosterone.

Oestrogen and progesterone have their strongest pigmentary impact on the sexual skin and in the areolae, and it may be that melanocytes in these areas are especially sensitive to the sex hormones. Alpha-MSH and beta-MSH, for instance, do not have a preferential effect on sexual skin. Furthermore, because the genitalia have one of the highest melanocyte counts in the body (even including those sites that are habitually exposed to sunlight) there may be an evolutionary explanation for the development of this phenomenon (see p. 192). Recent work has shown that many patients with melanoma have oestrogen receptors in their tumours, although the role of these is undetermined. It is still not clear how the sex steroids enhance pigmentation, although it is possible that they interact with MSH peptides. In certain mammals ovarian hormones have been shown to stimulate the secretion of MSH by the pituitary (Friedmann & Thody, 1986).

Modifications in hormone balance
Pregnancy

One of the early signs of pregnancy is darkening of the nipples and areolae and, to a lesser extent, of the face, anterior abdominal wall and genitalia. These changes are known collectively as the *chloasma* (or *melasma*) of pregnancy and they increase as the pregnancy advances. During pregnancy there is a great increase in melanin formation by the epidermal melanocytes, and melanocyte counts are higher than in non-pregnant women of the same age group (Snell & Bischitz, 1963). In recently delivered women (3–11 months post-partum) the areolae were significantly darker than in age-matched women of similar parity (Garn & French, 1963). The greater the number of pregnancies the darker the areolae, indicating a cumulative retention of pigment. It is interesting that hair colour does not seem to be influenced by pregnancy, and this again demonstrates the functional dissociation between melanocytes in the epidermis and those in hair follicles.

Increased 'melanocyte-stimulating activity' has been found in the blood and urine during pregnancy but it is known neither which MSH peptides are responsible for this activity nor what is the origin of such peptides. Possible sources (other than the mother's pituitary) are the placenta and the foetal pituitary (Friedmann & Thody, 1986).

Oral contraceptives

The use of oral contraceptives has been associated with the development of a discoloration of the cheeks, forehead and nose similar to the chloasma of pregnancy (Fig. 2.2). In one study (Resnick, 1967) nearly 30 per cent of women manifested chloasma as a direct result of these agents. Facial hyperpigmentation did not completely regress after cessation of the contraceptive therapy; and it was noteworthy that 87 per cent of those with the condition had also developed chloasma during pregnancy, thereby providing some basis for predicting susceptibility. Most of the patients linked the darkening with prolonged exposure to sunlight.

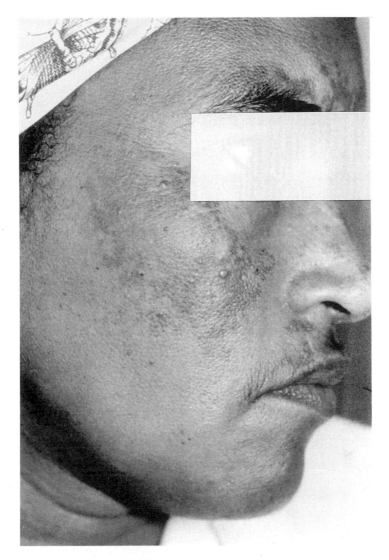

Fig. 2.2 Woman with chloasma due to prolonged treatment with oral contraceptive agents. Note blotches of hyperpigmentation on cheeks, nose and forehead. (Courtesy of Dr Jacques Cilliers.)

Microscopic examination of the epidermis revealed increased melano-genesis and the presence of enlarged melanocytes. Whether the cause of these changes resided in the oestrogen or the progesterone component of the contraceptive was unresolved. An Australian survey (Carruthers,

1966) found that the incidence of chloasma increased from 4 per cent during the first year of use to 37 per cent after 5 years of use. Exposure to sunlight was believed to act as a stimulus for melanocytes already primed by the sex hormones. Measurements of levels of 'immunoreactive beta-MSH' (now regarded as beta-LPH) in patients with this type of chloasma did not differ from those of unaffected age-matched controls (Smith *et al.*, 1977b). It seems therefore that the chloasma represents a specific sensitivity of facial melanocytes to the oestrogen and progesterone contained in oral contraceptives.

Menstrual cycle

There is some evidence that, like pregnancy, a similar but less marked chloasma occurs during the menstrual cycle. A questionnaire survey of female medical students and doctors (McGuinness, 1961) disclosed that nearly half of them were aware of fluctuations in skin pigmentation within the cycle – an increase in the later days and sometimes even during menstruation. Characteristically, darkening appeared around the eyes and mouth and in the areolar skin. Another study (Wassermann, 1974) revealed that about 75 per cent of student nurses responding to a questionnaire noted premenstrual darkening (mostly periocular) – and the latter was particularly associated with those subjects who had experienced premenstrual breast discomfort.

Snell & Turner (1966) investigated 29 women, of whom 18 (and especially the dark-skinned brunettes) noticed darkening around the eyes immediately prior to menstruation. Melanocyte counts from abdominal skin showed no changes during the cycle, and skin reflectances showed little change except for a tendency for the cheek and lower eyelids to darken in the later part of the cycle. A reflectance study of breast pigmentation (Pawson & Petrakis, 1975) showed that the areolae were darker not only in pregnant women but also in non-pregnant women during the last week of the menstrual cycle.

Although menstrual chloasma is not an established entity, the above evidence indicates that some skin discoloration does occur in the premenstrual phase. It is probably more likely that this phenomenon is vascular in origin rather than being melanin-induced. The skin blood flow is slow early in the cycle but increases greatly as menstruation approaches (Edwards & Duntley, 1949).

Notes

1 A third group of pigments, the trichochromes, are related to the phaeomelanins and occur also in yellow or red hair and feathers.

2 A recent study of Japanese subjects revealed that, although eumelanin was the dominant pigment within their *epidermal* melanocytes, phaeomelanin was also synthesized there in the form of spherical melanosomes. (Nakagawa, H. *et al.* (1989). *Journal of Investigative Dermatology,* **92**, 488.)

3 Aaron Lerner of Yale University Medical School has dedicated himself to the study of the pigment cell for 40 years. His success in developing the endocrinology of melanin pigmentation has been remarkable, he and his team being the first to isolate both MSH and melatonin.

4 There is evidence that exposure to bright light may also reset disturbances in circadian clocks such as occur in jet lag (*Science* (1989), **244**, 1256–7.)

3 *Ultraviolet radiation and the pigmentary system*

The previous chapter discussed the variable influences of hormones on the epidermal melanin unit. The secretion of the pineal hormone, melatonin, is governed by the amount of light reaching the eye and this mechanism enables animals like the weasel and arctic fox to alter their coat colour according to the seasons. In humans light also plays a dominant role in pigmentation; not indirectly through hormones, but by the direct effect of solar ultraviolet radiation (UV) on the epidermal melanin unit. This effect induces the so-called *tanning reaction*, which can increase the pigmentation of sun-exposed areas markedly above the level of natural pigmentation. The contrast can be striking and, as Noel Coward observed, 'Sunburn is very becoming – but only when it is even – one must be careful not to look like a mixed grill'!

In this century, and particularly in the Western world, the pursuit of a tan has become a passion, and there are people who will spend hours sunbathing (Fig. 3.1) or in sunbeds and suntan parlours. The achievement of a bronzed appearance is believed to signify health and beauty whereas, in fact, exposure to the sun (particularly in vulnerable, light-skinned Caucasoids) can have the very harmful effects which are discussed below.

Types of ultraviolet radiation (UV)

UV is part of the electromagnetic spectrum and it lies between the visible and X-ray regions (Fig. 3.2). Of the total radiant energy received by the earth from the sun, only 5–10 per cent is in the ultraviolet, the remainder being divided between the visible (about 40 per cent) and the infrared (about 50 per cent). Different wavebands in the UV spectrum show different capacities to cause biological injury. It is thus useful to subdivide UV into three segments, and the following classification will be used in this text:

UV-A: wavelengths 320–400 nm
UV-B: wavelengths 280–320 nm
UV-C: wavelengths 200–280 nm

Fig. 3.1 Browning of fair-skinned Caucasoids. Determined tan-seekers expose themselves to strong ultraviolet rays of midday summer sun. Note marked degree of tanning achieved in some sunbathers. (Courtesy of *The Argus*, Cape Town.)

UV-C will not be mentioned further because wavelengths below 280 nm do not normally reach the earth's surface as they are screened out by the ozone layer in the stratosphere.

Effects of UV-B

This is the noxious component of terrestrial radiation and it is responsible *par excellence* for causing the *sunburn reaction*. The latter takes the form of erythema (reddening of the skin), which is due to the dilatation of the blood vessels in the dermis with an increased blood content. In addition to this redness there may be swelling of the skin (oedema), pain and

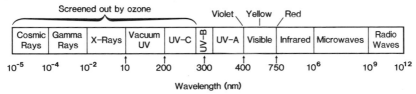

Fig. 3.2 Electromagnetic spectrum (not drawn to scale) with special reference to ultraviolet and visible regions.

blistering. UV-B not only causes erythema but also stimulates the production of the melanin which constitutes the basis for tanning.

Although only 1 per cent of the UV reaching the earth is UV-B the potential of UV-B to induce erythema is so great that it has sometimes been designated as the 'sunburn spectrum'. (Wavelengths longer than 320 nm are relatively inefficient at causing erythema.)

The UV-B portion of the spectrum can promote skin cancer, especially if exposure to it has been repeated and prolonged. The most common skin cancers in Caucasoids occur on the sun-exposed areas of the body (especially the head and neck). The incidence of skin cancer in Caucasoids approximately doubles for every 10 degrees decrease in latitude. UV-B intensity increases with decreasing latitude and reaches its peak at the equator (0°). Thus fair-skinned people in tropical and subtropical regions who have outdoor occupations (e.g. farmers, fishermen) are particularly at risk for skin cancer (see Fig. 9.4).

In addition to its cancer-producing effects, UV-B accelerates the degenerative skin changes associated with aging, eventually giving rise to a dry, leathery and wrinkled appearance. UV-B also has damaging effects on the eye: it is absorbed mainly in the superficial layers and causes conjunctivitis and corneal inflammation (keratitis).

The only beneficial effect of UV-B is that it stimulates the formation of vitamin D in the epidermis. Vitamin D promotes the absorption of calcium from the intestine and ensures the proper mineralization of bone. Deficiency of the vitamin was once a frequent cause of rickets in children, especially those deprived of sunlight in the dark, smoke-polluted areas of northern industrial cities (see Chapter 11 for a detailed discussion of this subject). Rickets has been virtually eliminated since the supplementation of foodstuffs with vitamin D. It must be emphasized that exposure of a small area of the body to a small amount of UV-B (about 5 per cent of that required for erythema) is all that is needed for adequate skin synthesis of the vitamin. There is therefore no justification for sunbathing in order to achieve a normal vitamin D status.

Effects of UV-A

This is sometimes referred to as 'long-wave ultraviolet' because at its longer end it impinges on the visible spectrum of light (see Fig. 3.2). The amount of UV-A reaching the earth's surface is several orders of magnitude greater than the amount of UV-B.

UV-A has weak erythema-producing effects compared with UV-B (1000 times less) and, although it does stimulate melanin pigmentation, the resultant tan appears to be more transient and less protective against

UV-induced injury than the one following exposure to UV-B. However, the acquisition of 'a tan without the burn' is appealing, and so the artificial application of UV-A (as in the UV-A sunbed) has been exploited to provide tan-seekers with a quick, effective and pain-free result, no doubt at some financial cost! Such a measure is to be avoided in those who either do not tan or tan poorly as these individuals are still likely to develop erythema. Under controlled conditions, UV-A has found a place in the treatment of psoriasis and vitiligo (see below). The presumed innocuousness of UV-A has recently been reappraised, and the profile that now emerges is not so benign. It is currently accepted that UV-A (especially after repeated or prolonged exposure) may cause erythema and skin damage and that it probably augments the deleterious effects of UV-B, including the risk of carcinogenicity. It is involved in adverse drug-induced reactions to light (photosensitivity) and in certain skin diseases (photodermatoses). Moreover, because it is absorbed in the lens of the eye, it may cause cataract.

Photoimmunology

Recent work has established a link between UV and immunity, and this is not surprising as the skin is an active immunological organ (*Lancet*, 1983, 1986). There is evidence that immunological factors play a part in the development of UV-B-induced skin cancers. Patients on drugs which suppress the immune system are more prone to skin cancers of the light-exposed areas. Mice which have received UV-B irradiation lose their ability to reject transplanted skin cancers. The explanation for this failure to reject tumours rests with a specific cell, the suppressor T lymphocyte. Normally, when the skin is invaded by foreign (antigenic) tissue, specific T-helper cells are evoked that migrate to the epidermis to counteract the antigen. UV-B seems not only to diminish the T-helper cell response but also to stimulate suppressor T lymphocytes which have the specific effect of reversing normal immune processes. This phenomenon provides a possible mechanism whereby UV induces skin cancer. The observation that recreational sunbathing may affect the immune system and possibly cause systemic immunosuppression makes photoimmunology an exciting avenue for future research.

Factors influencing ultraviolet radiation

The total solar radiation reaching the surface of the earth is made up of two approximately equal components – a direct component (sunlight) and a diffuse component which is scattered by the sky (skylight).

Although the total amount of UV from skylight equals that from sunlight, the skylight component may actually be greater at certain times of the day (early morning or late afternoon) or under certain conditions. It is possible, for example, to be sunburnt on the beach (an area receiving diffuse skylight from the large and open expanse of sky) even though shielded by an umbrella. It is interesting that, of various outdoor vacation activities, sunbathing on the beach scored the highest in amount of ambient UV-B received – 75 per cent, compared with 22 per cent for skiing and 14 per cent for sailing (Diffey, Larko & Swanbeck, 1982).

Ozone

This is a strong absorber of radiation between 220 nm and 300 nm. It is concentrated in the stratosphere, 10–50 kilometres above sea level, and the total amount present is equal to a layer 3 millimetres thick at normal temperature and pressure. Its concentration increases with latitude (being 50 per cent greater at the poles than at the equator) and varies with the seasons (being greatest in late winter and early spring).

An alarming development in recent years has been a depletion of atmospheric ozone due to the discharge of chemicals (chlorofluorocarbons) from sources such as aerosol propellants, refrigerants and effluents from supersonic aircraft engines. If this trend continues then ozone depletion could be as much as 6–10 per cent or more by early in the twenty-first century. This is very serious from the public health standpoint: it is estimated that for every 1 per cent decrease in ozone there will be a 2 per cent increase in the UV reaching the earth and a 4–6 per cent increase in the incidence of common skin cancers. Furthermore, since the 1970s there has been an alarming decline (about 40 per cent) in springtime ozone levels over Anarctica (the Antarctic 'hole') and a less marked (but disturbing) decrease over the North Pole. The consequences of ozone-layer thinning extend not only to skin cancer[1] (and cataracts) but to food chains, crops and climate. The global implications are such that at Montreal in September 1987 a United Nations treaty was concluded which aims to curb the worldwide production of chlorofluorocarbons by 50 per cent during the 1990s.

Altitude

In general, for every 500 metres ascent there is a 5–10 per cent increase in the sunburn effectiveness of UV, i.e. at an altitude of 2 kilometres the intensity of UV-B is about one-third greater than at sea-level. Conversely, at the Dead Sea (which is the lowest place on earth at 400 metres below sea level) the lowest levels of UV-B are encountered – between

11.00 and 13.00 in mid-summer the UV-B intensity there is the same as that at Beersheba (280 metres above sea level) at 19.00 in mid-summer (Kushelevsky & Slifkin, 1975). Because the UV-B rays are filtered out, sunburn is much less of a problem there than elsewhere. Moreover, there is a relatively greater exposure to UV-A and this unique radiation profile has made the Dead Sea into a centre for the treatment of psoriasis. Patients are able to remain in the sun (in an enchanting setting) for extended periods while they receive the UV-A that would otherwise have to be delivered artificially.

Clouds, humidity and surface reflection

The presence of clouds can reduce UV, particularly when the sun is obscured, but because of their water content clouds reduce infrared radiation (heat) to a greater extent than they reduce UV. This creates the erroneous impression that because it is a cool, cloudy day there is little possibility of sunburn. Moreover, experiments with hairless mice have shown that high humidity and water immersion enhance UV skin injury (Owens *et al.*, 1975). Atmospheric pollution by dust and smoke may eliminate much of the UV destined for the earth.

Surface reflection is also an important determinant of the amount of UV-B received by an individual. Overall, planet earth reflects about 35 per cent of the solar radiation striking it. In terms of UV-B, snow reflects 85 per cent, a grassy area 3 per cent, the ocean 3 per cent and sand about 15 per cent. UV injury is therefore common in snowfields and on snow-capped mountains where sunburn as well as corneal injury ('snow blindness') are sustained. For the prevention of the latter, skiers and mountaineers wear sun-glasses. The discovery of wooden slit goggles in Arctic middens shows that the early inhabitants of these regions had a need to protect against snow blindness.

Daily, latitudinal and seasonal variation

Most of these variations are self-evident. The intensity of UV increases and decreases according to the altitude of the sun, being very small in the early morning and later afternoon and maximal around noon when the sun reaches its zenith. On average, about two-thirds of the total daily UV-B intensity is received by the earth between 10.00 and 14.00.

There is a considerable variation in UV influx with latitude, the shorter wavelengths showing a much sharper variation owing to their strong absorption by ozone. For example, for 290 nm the annual influx at the equator is nearly 300 times greater than that at latitude 50° N whereas for

340 nm there is less than twice the difference between the two locations (Johnson, Mo & Green, 1976).

With regard to the time of year, at the higher latitudes the summer months provide the major portion of the total annual UV energy, with the shorter wavelengths contributing disproportionately much more than the longer wavelengths. At 50° N, for example, 290 nm radiation is 2000 times more intense but 340 nm radiation is only 10 times more intense in the month of July than in January (Johnson *et al.*, 1976). At the equator, on the other hand, there is hardly any difference in the intensities of either 290 nm or 340 nm radiation in July as opposed to January.

The effect of ultraviolet radiation on the human melanocyte system

Darkening of the human skin after exposure to the sun or to artificial UV involves two discrete stages: *immediate tanning* and *delayed tanning*, each characterized by a distinctive microscopic appearance (Jimbow *et al.*, 1974).

Immediate tanning

This causes the transient brownish tan which follows exposure to UV-A and visible light (320–700 nm radiation). It begins promptly after exposure, reaches a maximum within 1–2 hours, and then fades between 3–24 hours after exposure. If the initial exposure has been prolonged the fading time will increase and the immediate tanning will overlap with delayed tanning.

Delayed tanning

This gives rise to the familiar, sought-after and durable tan, which is induced by repeated exposure principally to UV-B but also to UV-A and to visible light. It is a gradual process in which the skin starts darkening 48–72 hours after irradiation. According to the reflectance studies of Edwards & Duntley (1939a) the tan reaches a maximum 19 days after an exposure and the skin does not return to its original melanin content until about 9.5 months afterwards. With repeated small exposures, however, maximal darkening is usually achieved between days 7 and 9 (Lee & Lasker, 1959).

Immediate tanning differs in its mechanism from delayed tanning. Certain ultrastructural changes have been observed in melanocytes in immediate tanning (e.g. the appearance of thick filaments and microtubules, and the translocation of melanosomes from the perinuclear area to the dendritic processes) but no actual increase in the size and number of melanosomes (Jimbow *et al.*, 1974). However, a recent study (Hönigs-mann *et al.*, 1986) found virtually no *structural* differences between the melanocytes and keratinocytes of skin showing immediate tanning and those of non-irradiated control skin. It thus seems probable that the immediate tanning reaction is based on a photo-oxidation of pre-existing melanin, melanin precursors, or even of other epidermal constituents.

Delayed tanning, on the other hand, is dependent on both qualitative and quantitative changes within the melanocytes. The latter enlarge in size, increase their dendritic density and develop a diffuse distribution of thick filaments in their cell bodies. There is an expansion of the ribosomes, endoplasmic reticulum and Golgi apparatus (reflecting an increase in the synthesis of tyrosinase and melanosomes), an increase in the numbers of melanosomes at all developmental stages, in their melanization and in the numbers transferred to keratinocytes. The numbers of functioning (dopa-positive) melanocytes also increase, although it is not known to what extent this phenomenon is due to the generation of new cells by mitosis and/or to the activation of dormant melanocytes.[2] Delayed tanning is therefore due to an increase in melanocyte numbers and in melanogenesis. A recent and novel hypothesis proposes that it is vitamin D (formed in the skin by UV-B) that stimulates the melanocytes and mediates the photo-induced pigmentary response (Tomita, Torinuki & Tagami, 1988).[3]

Exposure to UV-B and UV-A is reported to have increased plasma and urinary levels of 5-*S*-cysteinyldopa and urinary 6-hydroxy-5-methoxyindole-2-carboxylic acid (see p. 31) (Rorsman & Tegner, 1988), but these changes were related to UV-induced pigmentation and especially to erythema. In particular, the raised urinary 5-*S*-cysteinyldopa was associated with UV skin damage and not with increased melanocyte activity (Stierner *et al.*, 1988), confirming the view that 5-*S*-cysteinyldopa in urine mainly originates outside the melanocyte.

Racial differences

The cellular changes after UV exposure are so specific that electron microscopy can differentiate between skin specimens derived from exposed and unexposed areas. After irradiation numerous stage 4 melanosomes appear in Caucasoid melanocytes, including those from red-haired and fair-skinned subjects whose unexposed skin usually contains very few stage 4 melanosomes. In Negroids (whose non-irradiated skin contains mainly stage 4 melanosomes), stage 2 and 3 melanosomes emerge following UV irradiation.

The racial pattern of melanosome distribution in keratinocytes (see p. 17) remains the same before and after irradiation, both in the immediate and in the delayed tanning reactions (Jimbow *et al.*, 1974). However, there is some conflict in this matter, and it may be that after excessive UV exposure the melanosomes in Caucasoids and Mongoloids increase in size to an extent that their typically aggregated configuration within keratinocytes alters to the single state characteristic of Negroids (Toda *et al.*, 1972).

Natural photoprotection

The tanning reaction to UV is a consequence of acute damage sustained by epidermal cells. Melanogenesis is stimulated in order to provide a protective melanin barrier within the keratinocytes. (Melanin is the natural sunscreen and its properties will be discussed in the next chapter.) It is the delayed tanning which provides the photoprotection, the immediate tanning being ineffective in this respect (Black, Matzinger & Gange, 1985).

Apart from melanin the following are other components within the epidermis which have the potential to protect against UV injury.

Stratum corneum

The thickness of the stratum corneum has a definite protective function. The mitotic rate of the basal keratinocytes increases a day after UV exposure, reaches a maximum two days later and maintains this level for about a week. It then declines, and provided that there has been no subsequent exposure the skin regains its original thickness after one to two months. The keratin and proteins within the stratum corneum act mainly by scattering and absorption of UV. The palms and soles are the regions with the thickest stratum corneum and these areas are exceptionally resistant to UV damage.

Urocanic acid and carotenoids

Urocanic acid, a deamination product of the amino acid histidine, is a normal constituent of the epidermis and, on exposure to radiation, it undergoes a chemical reaction which absorbs the UV. It is, however, considered relatively weak as a natural sunscreen against UV-B (Pathak & Fitzpatrick, 1974). Carotenoids are normally present in the human epidermis and dermis in extremely small quantities, and they can only protect the skin if they are administered (in the form of beta-carotene) in high oral doses. They operate chiefly within the visible spectrum, and they therefore have a place in protecting those patients who develop photosensitivity reactions to visible light (Pathak & Fitzpatrick, 1974). They are ineffective in preventing UV-induced skin damage (Wolf, Steiner & Hönigsmann, 1988).

Sunscreen preparations

Many pale-skinned Caucasoids fail to acquire the tan they so eagerly desire. Instead they are highly vulnerable to sunburn, a tendency that is a fairly reliable predictor of skin cancers in later years. Such people must take daily precautions to protect their skins. This is particularly important in children because, although skin cancer usually declares itself in middle life, it often has its beginning in early childhood.[4] Of course, the most practical and indeed the universal form of sunscreening is the use of clothing and headgear; but for those who seek naked exposure sunscreen agents offer reasonable photoprotection provided they are applied with an understanding of their properties and limitations. They are not recommended for babies or infants because of a lack of data on the absorption and metabolism of sunscreen ingredients in very young children.

Types of sunscreen agents

A sunscreen preparation is applied topically to reduce the intensity of UV impinging on the skin. Its ability to screen out UV must be coupled with several essential physical features – it must be non-irritant, non-absorbent, stable to heat and light, easily and effectively applied, and cosmetically acceptable. One of its main attributes is perhaps its *substantivity*, i.e. its resistance to being removed by swimming or sweating, and this depends on its physico-chemical adhesiveness to the stratum corneum.

Physical agents

Sunscreens are usually either physical or chemical. The physical agents usually contain opaque ingredients, zinc oxide or titanium dioxide, and they form a visible film on the skin surface which blocks the transmission of UV-B, UB-A and visible light to the underlying epidermis. This makes them broad-spectrum sunscreens. They are cosmetically unattractive, and they interfere with perspiration so that the skin feels unpleasantly hot. Their use is generally restricted to small and vulnerable areas such as the lips and nasal bridge.

Chemical agents

The chemical preparations are the sunscreens of choice because they are invisible on the skin after application. The commercially available chemical sunscreens belong to four main groups: derivatives of *para*-aminobenzoic acid (PABA) and its esters, benzophenones, cinnamates and salicylates. All of these, except the benzophenones, absorb exclusively in the UV-B (sunburn) range. PABA and its esters are probably the most effective and best-investigated compounds in the protection against UV-B sunburn. Benzophenones absorb UV-B (although less efficiently than the PABA group) and they have the additional advantage of absorbing UV-A up to 360 nm. Butylmethoxydibenzoylmethane (BMDM) absorbs in the range 320–390 nm. The most appropriate chemical sunscreens combine a benzophenone and/or BMDM with a PABA derivative, cinnamate or another of the UV-B absorbers.

It is important to explode a myth which is disseminated by various advertisers, namely that certain sunscreens stimulate melanogenesis and therefore promote tanning. This is fallacious, because if tanning does occur after sunscreen applications it is not due to the action of the sunscreen. With a preparation containing PABA, the UV-B is eliminated and on repeated exposures UV-A itself will lead to tanning. Therefore, tanning in this situation is not because of the sunscreen but in spite of it! In individuals who have an inherent capacity to tan the use of a UV-B blocker will allow tanning from the UV-A with little resultant erythema, thereby encouraging prolonged sunbathing and excessive UV-A intake. As mentioned earlier, prolonged and multiple exposures to UV-A may not be safe. An hour in the sun every day for 2–3 months a year over 4 consecutive years would result in an accumulated total dosage of UV-A which is severely damaging to the skin, especially the dermis (Kligman, Akin & Kligman, 1985). This potential risk compels the use of a sunscreen which also contains a UV-A filter (e.g. BMDM). The danger is compounded in certain photosensitivity conditions (see below) which are triggered by UV-A. As many of the available sunscreens protect against UV-B only, their use results in over-exposure to UV-A and precipitation of the photosensitivity disorder.

Skin darkening agents

Melanocyte-stimulating hormone (MSH) is able to darken human skin after intramuscular injection (see p. 32). Daily injections of MSH will increase the darkening until the injections are discontinued, after which the skin will return to its original colour in about a month. The need for regular and frequent injections of MSH to maintain the skin darkening effect limits the practical usefulness of this compound.

More reliable and convenient agents for inducing skin darkening are the *psoralens*. These drugs are photosensitizers (see below) and their oral ingestion, followed about 2 hours later by exposure to sunlight or to UV-A, leads to an augmented erythemal response. Graduated and prolonged psoralen treatment will produce both a marked hyperpigmentation and an increased thickness of the stratum corneum. The therapy will therefore enhance tanning capacity and improve tolerance to the sun (i.e. reduce susceptibility to sunburn). The psoralens appear to act by undergoing a light-dependent conjugation with the DNA of melanocytes, with a consequent increase in the number and size of the latter and in the activity of tyrosinase. The result is an accelerated formation, melanization and transfer (to the keratinocytes) of melanosomes.

Therapy with psoralens and UV-A (known by the acronym PUVA) has been successfully employed in psoriasis (in which it appears to inhibit epidermal cell division)[5] and in vitiligo (in which it induces fairly persistent repigmentation of the depigmented skin areas). However, a recent sophistication by the pharmaceutical industry has been the addition of psoralen to a sunscreen preparation in order to promote tanning and sun tolerance. There is evidence that such psoralen-containing sunscreens do induce a tan (even in subjects with otherwise poor tanning capacity) which provides protection against the damaging effects of solar UV (Young *et al.*, 1988). However, concern has been expressed that psoralens may cause cancer in the long term,[6] possibly by bonding to DNA (*Lancet*, 1981), and their use in sunscreens best awaits further research into the safety aspects.

There are certain compounds which produce simulated ('out-of-the-bottle') tanning. The best known of these is *dihydroxyacetone*, which chemically reacts and binds with proteins in the stratum corneum to form an orange-brown colour. It actually dyes the stratum corneum but the resultant coloration does not screen out UV-A or UV-B and, like other agents with a similar effect, dihydroxyacetone is ineffective as a sunscreen.

Another agent, *canthaxanthin*, has been promoted as a 'quick suntan' pill. Canthaxanthin is a carotenoid naturally present in crustaceans, fish, mushrooms and yellow and orange vegetables. It can be synthesized and it is used as a food colouring agent in butter and cheese. Given in adequate dosage, it is deposited in body fat and causes a 'tan', although sometimes the carrot-like, orange-brown colour is unsightly. Furthermore, it affords no protection against the harmful effects of the sun.

Canthaxanthin is also associated with gold-coloured crystalline deposits in the retina which may reduce vision at low light intensities and impair dark adaptation. The risk is not considered to be warranted for cosmetic indications, and these 'suntan' canthaxanthin products have been withdrawn from the market in several countries. Canthaxanthin (combined with beta-carotene) has a justifiable place in the treatment of a certain photosensitive skin condition (see below).

Finally, high-dosage vitamin A preparations have been administered for the prevention of sunburn. Not only are they useless for this purpose but, in view of the potential toxicity of vitamin A, they may be dangerous. Moreover, in combination with tetracycline antibiotics, they can lead to serious elevations in intracranial pressure.

Sun protection factor

There is a wide assortment of sunscreen preparations available on the market. One of the problems in the past has been to provide a scientific framework by which the efficacy of these preparations can be quantified and standardized. Such a system has now been developed. It is based on the concept of the *minimal erythemal dose* (MED): this is defined as the lowest dose of UV of a specified waveband (UV-A or UV-B) which evokes a minimally perceptible redness 24 hours following exposure.

An index known as the *sun protection factor* (SPF) is derived for any given sunscreen by the following calculation:

$$\frac{\text{MED of sunscreen-protected skin}}{\text{MED of unprotected skin}}$$

In practical terms, it is generally agreed that 15 minutes of exposure to the midday sun in mid-summer will elicit a minimal perceptible redness 24 hours later. Thus, in a sense, the MED of the midday summer sun is 15 minutes of exposure. An applied sunscreen with an SPF of 4, for example, would enable a person to tolerate 1 hour of exposure under identical conditions before minimal erythema appears. Exposure to 10 to 14 times the MED (i.e. 2.5–3.5 hours of exposure) would cause severe sunburn with oedema and blistering. In theory, a sunscreen with an SPF of 10 to 15 should provide adequate protection against this hazard.

The SPF has been widely used in Europe and America as a measure of sun-barrier efficiency. An SPF of 2 would give minimal protection and one of 15 almost full protection. The United States Food and Drug Administration requires SPF testing for all sunscreens, which are now deemed to be 'drugs' (and not cosmetics) with a specific place in the prevention of skin disease.

Kaidbey (1990) has proposed that so-called 'sunburn cells' (keratino-cytes with microscopic evidence of degeneration) are an earlier and more sensitive index of UV damage to DNA than erythema and that, to prevent the appearance of these cells after solar exposure, the new 'superpotent' sunscreens (SPF of 30) are significantly more effective than the SPF-15 products.

Skin types and screening effect

On the basis of an individual's skin response to the sun, it is possible to define different skin types. A convenient classification is shown in Table 3.1, together with the recommended SPF for a particular skin type as determined from outdoor field studies (Pathak, 1982).[7] It is suggested that individuals of skin types I and II should use a sunscreen with an SPF of 10 or more regardless of the intended length of exposure. Their susceptibility to premature skin aging and skin cancer warrants the application of a sunscreen as a regular part of the daily toilet.

Kidney transplant patients are another group requiring sunscreens. These patients receive immunosuppressive therapy and, in view of the additional immunosuppressive effects of UV-B (see above), they are at risk of developing skin cancers and other sun-related skin lesions (Boyle *et al.*, 1984).

Photosensitivity disorders

Skin photosensitivity is an abnormal reaction of the skin to light, usually mediated by the action of certain drugs or chemicals (photosensitizers). The latter may either be administered orally or applied topically, and they include medicines such as the phenothiazines (used in the treatment of schizophrenia) and the sulphonamide and tetracycline antibiotics. As

Table 3.1 *Relationship between skin type, response to solar exposure and recommended SPF of sunscreen (from Pathak, 1982)*

Skin type	Response	Recommended SPF
I	Always burns easily; never tans	10 or more
II	Always burns easily; tans minimally	10 or more
III	Burns moderately; tans gradually and uniformly (light brown complexion)	8–10
IV	Burns minimally; always tans well (moderate brown complexion)	6–8
V	Rarely burns; tans profusely (dark brown complexion)	4
VI	Never burns; deeply pigmented (black)	None indicated

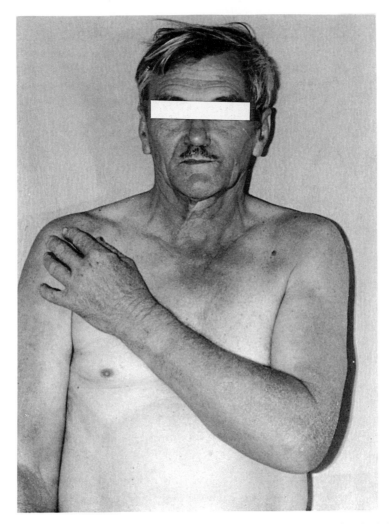

Fig. 3.3 Drug-induced skin photosensitivity. Sun-exposed areas (particularly hands and arms) show marked erythematous reaction (redness and swelling) reminiscent of exaggerated sunburn. (Courtesy of Professor Norma Saxe.)

mentioned above, the psoralen agents are photosensitizers whose properties are specifically harnessed for therapeutic use in dermatology. In themselves photosensitizing agents are harmless to the skin, but on exposure to UV (usually UV-A) they precipitate an acute response known as a *phototoxic reaction*: this clinically resembles an exaggerated sunburn (with signs such as painful swelling, blisters and urticaria) and is confined to the sun-exposed sites (Fig. 3.3). A less common reaction, known as a *photoallergic reaction*, occurs when UV (also usually UV-A) induces a photochemical interaction between a photosensitizing drug and skin protein, resulting in the formation of an antigen. This antigen then triggers an immunological hypersensitivity reaction (such as urticaria or eczema) in which the skin eruptions extend beyond the exposed areas.

In a group of metabolic disorders known as the *porphyrias* the photosensitivity reaction is due to an overproduction in the body of porphyrin compounds and their precursors. These endogenous porphyrins are potent photosensitizers and, on exposure to light (usually visible light, and particularly the wavelengths 400–410 nm), they provoke the typical features of burning, itching, swelling and blistering – in severe cases, the skin can become scarred and disfigured. The most disabling of the porphyrias are the very rare genetic diseases, erythropoietic porphyria and erythropoietic protoporphyria. In the latter condition the oral ingestion of beta-carotene (alone or combined with canthaxanthin) relieves the photosensitivity and enables patients to tolerate sunlight exposure without discomfort. The beta-carotene probably acts by quenching the noxious and reactive substances liberated in the pathogenic photochemical reaction specific to erythropoietic protoporphyria. Beta-carotene has little value in any other photosensitivity states.

Finally, there are skin disorders which exhibit abnormal light reactions without the participation of any known photosensitizer. The most common of these (and possibly the commonest photosensitivity disorder of all) is *polymorphous light eruption*. This manifests as redness, papules, blisters and eczema, all usually confined to the exposed areas. The lesions are seasonal – typically appearing in the spring and summer – and they are induced chiefly by UV-A.

The prevention of photosensitivity reactions lies primarily in the avoidance of sunlight. Sunscreens which contain only UV-B blockers are potentially dangerous because they encourage sunbathing with excessive exposure to UV-A which is the harmful component. In the porphyrias, where visible light is the culprit, only the physical sunscreens (titanium dioxide or zinc oxide) are likely to be protective.

Notes

1 The United States Environmental Protection Agency has estimated that a 50 per cent global loss of ozone by the year AD 2050 would result in an extra 160 million cases of non-melanoma skin cancer in the United States population born between AD 2030 and AD 2074.

2 Repeated UV irradiation of mouse skin caused a four-to-six-fold elevation in the melanocyte population due primarily to mitosis (Rosdahl, I.K. & Szabo, G. (1978). *Journal of Investigative Dermatology, 70*, 143–8).

3 UV exposure has resulted in an increase in circulating levels of pituitary pigmentary hormones (e.g. MSH, beta-LPH) in animals and humans, and another suggestion for the mechanism of photo-induced pigmentation is that UV-B induces melanogenesis by increasing MSH-receptor activity on cutaneous melanocytes (Bolognia, J., Murray, M. & Pawelek, J. (1989). *Journal of Investigative Dermatology, 92*, 651–6).

4 About 80 per cent of skin damage occurs within the first 20 years of life.

5 Recently, an active metabolite of vitamin D_3, administered orally and topically, has been shown to cause significant improvement to patients with psoriasis, possibly by inhibiting proliferation and inducing differentiation of epidermal cells (Holick, Smith & Pincus, 1987).

6 The latest work reports that PUVA has carcinogenic properties. Men with psoriasis who were exposed to prolonged PUVA therapy showed a markedly increased susceptibility to skin cancer which occurred particularly on the genitalia (Epstein, J.H. (1990). *New England Journal of Medicine, 322*, 1149–51).

7 Recent evidence throws some doubt on the usefulness of skin type as a predictor of the responses of the skin to UV irradiation, and objectively measured constitutional skin colour appears to be more reliable in this respect (Westerhof, W. *et al.* (1990). *Journal of Investigative Dermatology, 94*, 812–16).

4   *Functions of melanin*

Photoprotection

The two major defences of the skin against radiation injury are the presence of melanin pigment and the thickness of the stratum corneum. There was a time when the role of melanin in photoprotection was subordinated to that of the stratum corneum but, in the recent past, the burden of evidence has declared melanin to be the natural sunscreen *par excellence*. The arguments in favour of the superior photoprotective properties of melanin have been convincingly set out by Pathak & Fitzpatrick (1974) and they are based on clinical, epidemiological and experimental findings.

Skin cancer

The most obvious clue to the photoprotective role of melanin resides in the prevalence of skin cancer, which is by far the commonest of the cancers. As already noted, it is associated with intense and long-term exposure to UV-B and it therefore occurs more frequently on the chronically exposed body areas, such as the head and neck (see Fig. 9.4). There is a relative infrequency of skin cancer in Negroids and other pigmented peoples (Amerindians, Asians), even at the equator where UV is strongest. Susceptibility to skin cancer (including malignant melanoma) is enhanced in fair-skinned, light-haired Caucasoids who sunburn easily and tan poorly.

Persons of Celtic background appear to be significantly over-represented in the skin cancer statistics (Urbach, 1969). The Republic of Ireland has the third-highest death rate from skin cancer (next to Australia and South Africa), even though it is located between 52° N and 54° N and receives a relatively low annual influx of UV-B. It may be that the individual with Celtic skin and red hair has a genetic inability to resist the deleterious effects of UV. Indeed, Ranadive *et al*. (1986) have shown that phaeomelanin from red hair, when irradiated with wavelengths 320–700 nm, causes considerably more damage to cells *in vitro* than did eumelanin from black hair. The incidence of skin cancer is from five to eight times more common in fair-skinned Caucasoids living in areas of high solar radiation (e.g. Texas and Arizona in the United States of America) than the average incidence for Caucasoid Americans. As already noted, it is estimated that in Caucasoids there is an approximate doubling of the incidence rate of skin cancer for every 10° decrease in latitude.

Although Negroids are very resistant to the cancer-inducing effects of UV, this generalization does not apply to the Negroid albino, who lacks melanin pigment. Negroid albinos are highly susceptible to sunlight (much more so than light-skinned Caucasoids), and they develop skin cancers and related skin lesions at a very early age (see p. 144 and Fig. 9.4). This observation constitutes a strong pointer to the specific photo-protective effect of melanin. Even in the experimental situation the data are consistent. When skin tumours are induced in mice by UV, albino mice develop a far higher incidence of tumours than brown or black strains and the tumour-induction period in albinos is about half of that in pigmented animals (Pathak & Fitzpatrick, 1974).

Thickness of the stratum corneum

If the stratum corneum, rather than melanin, is the major determinant of the skin's photoprotective armoury then one would have to argue that pigmented people are resistant to UV-B because they have a greater thickness of this layer.

The matter was investigated in a classic study by Thomson (1955) who obtained specimens of stratum corneum from the unexposed skin of Negroid and Caucasoid subjects. Although there was a wide variation in the thickness of the stratum corneum (especially in the Negroid group), there was no significant difference between the mean thickness in Negroids (11.13 μm) and that in Caucasoids (9.51 μm). A later study (Weigand, Haygood & Gaylor, 1974) showed that, although the average thickness was almost the same in Negroid and Caucasoid specimens (6.5 μm and 7.2 μm respectively), the former had more cell layers and a greater cell density than Caucasoids, suggesting a more compact stratum corneum.

Transmission of UV through skin

Thomson (1955) found that the transmission of solar UV (300–400 nm) was 3.5 times greater in Caucasoid (64 per cent) than in Negroid (18 per cent) stratum corneum. Skin pigment was held to be the most important factor responsible for this discrepancy. Among the specimens of stratum corneum was one from a Negroid albino, with a transmission value which fell totally outside the Negroid but within the Caucasoid range.

More sophisticated spectroscopic studies have been conducted on both the stratum corneum and the whole epidermis. Fair-skinned Caucasoid stratum corneum transmitted 30–35 per cent of UV at 300 nm and 65–70 per cent at 400 nm. The figures for whole epidermis (stratum corneum plus underlying layers) were 10–15 per cent at 300 nm and 50–55 per cent at 400 nm. These contrasted with 2–5 per cent and 12–20 per cent respectively for the intact epidermis of Negroid subjects (Pathak & Fitzpatrick, 1974), a difference that was due to the absorption and scattering of the incident radiation by melanin.

The work of Kaidbey *et al.* (1979) incorporated a specific analysis of transmission first through stratum corneum alone and then through the underlying epidermal layers. Although Caucasoids had five times more UV-A and UV-B transmission to the upper dermis than Negroids, the main site of UV filtration (particularly UV-B) in Caucasoids was the stratum corneum, whereas in Negroids it was the epidermis beneath the stratum corneum. In fact, transmission of UV-B was only 1.5 times more in Caucasoid than in Negroid stratum corneum – hardly a striking difference. This observation further devalues the role of stratum corneum in photoprotection; it suggests that, by the time melanin in black skin arrives at the stratum corneum, it has lost much of its capacity to absorb and scatter radiation. In Negroids the cell layers beneath the stratum corneum contain numerous and large melanosomes, singly dispersed and shielding the nuclei (see p. 17): these characteristics provide a dense and very effective UV filter.

Stratum corneum stripping

The ascendancy of the stratum corneum in photoprotection received its death-blow from an experiment in a subject with vitiligo, a condition characterized by patches of depigmentation. The horny cells of the stratum corneum were stripped by adhesive tape from both the pigmented and vitiliginous areas, and these sites were then exposed to the mid-summer sun for 45 minutes and 90 minutes. After 24 hours and 48 hours the stripped pigmented skin showed a very faint redness whereas the stripped depigmented skin developed a fiery-red erythema with oedema and blisters (Pathak & Fitzpatrick, 1974). The only difference between these two adjacent areas in the same individual was the presence and absence respectively of melanin – *quod erat demonstrandum*.

Melanin as a sunscreen

A similar experiment to that described above has demonstrated that, in pigmented patients with vitiligo, 45 and 90 minutes of sunlight exposure in mid-summer produced marked erythema in the vitiliginous area and negligible erythema in the adjacent pigmented skin (Pathak & Fitzpatrick, 1974).

Sun exposure of this degree is equivalent to three and six times the MED (minimal erythemal dose) (see p. 54), a dose range which the skin melanin easily withstands. Attempts have been made to quantify the sunscreening ability of natural melanin. Olson *et al.* (1973) compared the MEDs in subjects of various skin shades. They showed that, with deepening skin colour, there was an increase in the size of melanosomes, in the proportion of singly dispersed melanosomes, and in the MED. The average MED of dark-skinned Negroids was 33 times greater than that of Caucasoids, and nine times and five-and-a-half times greater respectively than that of light-complexioned Negroids and medium-complexioned Negroes. Calculations of the protection factor against UV-B revealed that for black epidermis it was 13.4, compared with 3.4 for white epidermis (Kaidbey *et al.*, 1979). The figure of 13.4 is matched by few commercially available sunscreens![1]

The free radical property of melanin

It has now been proved beyond doubt that melanin is a highly effective sunscreening compound and that it is the body's chief protection against solar radiation. The next question to consider is the mechanism by which it so ably exercises this function. The short answer is that it does so through *free radicals*.

Free radicals are chemical species with one or more unpaired electrons. They occur in plant and animal tissues and they are involved in metabolic activity and in various photochemical reactions such as oxidation-reduction and the action of UV and visible radiation on biological systems. Free radicals are often highly reactive, unstable and short-lived. An instrument called an electron spin resonance spectrophotometer can be used to estimate the *electron spin resonance* (ESR) of different materials by plotting signals relating to changes in absorbed energy. The ESR is an indirect measure of the free radical content of a material.

Melanin protects against damaging UV by scattering, by absorption of radiant energy, and by the dissipation of this energy as heat. There is also another mechanism available to it. Melanin itself can be thought of as a stable free radical: it contains within its polymeric structure units of 5,6-

dihydroxyindole in the semiquinonoid form (i.e. with unpaired electrons, and therefore free radicals).

Mason, Ingram & Allen (1960) found that the free radical content of human hair is proportional to its melanin concentration, being highest in black and dark brown hair. After irradiation, black hair shows a strong increase in free radical content while melanin-free hair gives a poor response. Sever, Cope & Polis (1962) showed that the melanin granules of the mammalian eye generate free radicals when irradiated with visible light, and Pathak & Stratton (1968) demonstrated that pigmented skin exhibits the stable ESR signal intrinsic to melanin, whereas 'white' skin does not. After exposure to UV-A and visible light this signal is enhanced in pigmented skin, but there is still no signal in non-pigmented skin. However, when both types of skin are irradiated with UV-B (290–320 nm), then a different (additional) ESR signal emerges, its contribution being almost 50 per cent greater in 'white' skin. This second signal appears to represent UV-generated reactive free radicals from epidermal proteins and other cellular constituents, but not from melanin.

The interpretation of these data is complex. It is speculated that, because of its own free radical property, melanin acts as a biological electron exchange or electron transfer polymer which participates in oxidation–reduction reactions and thereby protects tissues from the damaging effects of free radicals. A simplistic way of explaining this is that the stable free radicals within the melanin structure are capable of 'soaking up' and neutralizing the highly reactive, noxious free radicals which are liberated in cells by stimuli such as UV. This attribute makes melanin a potentially excellent electron acceptor, and it helps to explain the propensity of melanin to bind strongly to various drugs (a subject that will be discussed in the next chapter).

There is a school of thought which believes that melanin is not all good, but also has its negative aspects. While it undoubtedly acts as a photoprotective agent in certain situations it may cause photosensitization in other circumstances. Indeed, there is experimental evidence that the presence of melanin in cells such as macrophages and red blood cells leads to cytotoxicity and cellular death after UV irradiation (particularly at the shorter wavelengths of around 300 nm) (Menon & Haberman, 1977).

Thermoregulation

It is a physical property of black to absorb heat and of white to reflect it. This is easily demonstrated after a black-and-white dog, say, has lain in the sun – its black patches feel hot and the adjacent white areas relatively cool. Similarly it has been estimated that Negroids absorb approximately

30–40 per cent more heat from solar radiation than do Caucasoids. A black skin protects the tissues from UV damage. How does it react to a solar heat load?

It is pertinent to review some of the work done in birds and animals before addressing the problem in humans. Hamilton & Heppner (1967) found that white zebra finches exposed to artificial sunlight used an average of 23 per cent less energy after they were dyed black. Lustick (1969) repeated the experiment and confirmed that radiant energy lowered oxygen consumption by 26 per cent in dark birds and by 6 per cent in white ones. These observations suggest that melanin acts as a heat absorber and thereby minimizes the necessity for the bird to generate endogenous energy. It has been calculated that by sunning themselves in artificial sunlight, roadrunners (desert birds) saved energy equivalent to 41 per cent of their standard metabolism (Ohmart & Lasiewski, 1971). A metabolic saving of this order from solar heat enables the birds to curtail their food consumption – a distinct advantage, especially during winter months when the desert food supply declines. When the roadrunner assumes a sunning posture, its normally concealed areas of black skin and black plumage become exposed.

The above studies confirm that dark pigmentation is a better absorber of solar energy than light pigmentation. But this concept immediately introduces a paradox, namely, why one finds black birds and black beetles in the desert and white birds and white animals in the Arctic. It may well be that thermoregulation has not been a major selective factor in the evolution of colour, and that forces such as concealment and social interaction are of greater significance. It may also be that animals adopt behavioural mechanisms to offset the effects of colour on thermoregulation: in the case of birds, for example, orientation of posture towards the sun and feather erection.

There are, however, further considerations which contribute to an understanding of the paradox. It has been shown that the penetrance of solar radiation through fur is greater in white than in dark coats and that greater heating therefore occurs at the skin surface in light-coloured as compared with dark-coloured animals (Hutchinson & Brown, 1969; Øritsland, 1971). In other words, a white coat is a more efficient heat trap than its dark counterpart because its reflective properties transmit the radiative energy inwards towards the skin. This fact may explain why the white coat colour of Arctic inhabitants is thermally adaptive. Furthermore, Walsberg, Campbell & King (1978) found that, in black and white pigeons, the inherent capacity of the dark plumage to acquire a greater radiative heat load than the white diminishes with increasing wind velocity (especially in birds whose feathers are erected). This observation means that greater wind speeds cool dark plumages (by convection) more efficiently than they do white ones: hence, the thermal advantage to black birds in a hot, windswept desert environment.

The paradox has also been investigated in the context of the Bedouin practice of selecting the black goat rather than the white as the domestic animal in the Negev and Sinai deserts (Finch *et al.*, 1980). The black goats studied gained more heat and evaporated more water (to keep cool) than did their white companions in the hot desert at midday. Water is apparently not a crucial selection factor, and indeed the black goat shares with the camel an ability to drink 35 per cent of its body weight in water in a few minutes. It needs to be watered only once every 4 days during the hottest time of the year. It seems that it is not water economy but energy economy that determines survival for the goat (Dmiel, Prevulotzky & Shkolnik, 1980). Because the black goat absorbs more solar radiation than the white it would achieve (like the roadrunner) a sufficient reduction in metabolic rate to equip it for survival during the food scarcity of the short, cold desert winter.

Humans cannot tolerate dehydration for a prolonged period. More-over, the human is endowed with two to four million eccrine sweat glands (see Fig. 1.1(*a*)) and has the most efficient evaporative cooling system. Sweating is the first-line response to a heat load, whether this is due to a high ambient temperature or to muscular activity (exercise). Sweating under conditions of heat and sustained work can approach the level of 2 litres per hour, and these water losses must be replaced to avoid progressive dehydration (Newman, 1970). Camels can survive over a 30 per cent weight loss of water but the human is in dire straits at 10 per cent and dead at 18–20 per cent. Furthermore, because humans (unlike other mammals) are unable to consume large quantities of water rapidly they have to drink frequently to prevent dehydration.

Ethnic differences in thermoregulation

The situation arises in which humans have the least capacity of all mammals to consume water and yet the greatest dependence on thermal sweating (Newman, 1970). Indeed, Benjamin Franklin reasoned that Negroids would be at a considerable disadvantage in the tropics because of their greater heat absorption. It is important to test Franklin's supposition and, as a corollary, to enquire whether Negroids have more numerous sweat glands than Caucasoids to offset their apparently increased thermal load.

Several studies have assessed the heat tolerance of Negroids and Caucasoids. Robinson *et al.* (1941) investigated Negroid and Caucasoid farmers in Mississippi during work in a hot, humid environment. Both groups were roughly comparable in terms of height, body weight and state of acclimatization. The Negroids sweated less and remained cooler than the Caucasoids. Baker (1958) exercised Caucasoids and Negroid soldiers (matched for body fat, weight and structure) for one hour under mild, hot-humid conditions and found that the latter performed better. Two other studies under hot-humid conditions, one comparing African Negroids (Nigerians) and Caucasoids (Ladell, 1964) and the other American Negroids and Caucasoids (Riggs & Sargent, 1964), showed that the Negroids tended to sweat less and to be better adapted to the heat. Similarly, Strydom & Wyndham (1963) used standard laboratory conditions (a portable field tent with controlled temperature, humidity and air velocity) to test various ethnic groups (Negroids, San (Bushmen), Australian Aborigines and Caucasoids) in different climatic regions. They found that ethnic group *per se* was not an important factor in heat tolerance, but that the greatest variation among individuals occurred in the degree of their heat acclimatization at the time of testing. (Acclimatization is the process of physiological adjustment to, and therefore tolerance of, heat over a period of time). Given that acclimatization had been achieved, there was no difference in the heat tolerance of the different ethnic groups except that Negroids and Australoids sweated less than Caucasoids. There is evidence that Asian Indians and Mongoloids also sweat less than Caucasoids, and this seems to suggest that peoples of European ancestry have a tendency towards a higher rate of sweating in the heat (Hanna & Baker, 1974). Thus, Benjamin Franklin's hypothesis is rejected.

With hot-dry desert heat, however, Baker (1958) found that when Negroid and Caucasoid soldiers walked naked in the desert sun, the former tolerated the heat load less efficiently. This finding, which has never been corroborated, may be due to a greater heat absorption by Negroids.

Sweat glands

The older literature suggested that tropical people had more sweat glands than Europeans. Many studies have estimated the numbers of functioning eccrine sweat glands in various ethnic groups and these are summarized by Knip (1977). There is a very wide variation among individuals of the same ethnic group but ethnic differences as such are not evident. In short, different ethnic groups studied by the same investigator with the same method and in the same environment yield similar results. Factors such as climatic differences and acclimatization probably account for the major variations observed in the number and activity of human sweat glands.

Physique

There is another dimension to thermoregulation and that is physique. There are two well-known zoological rules relating to body size and climate. *Bergmann's rule* states that in colder climates body size increases; and *Allen's rule* states that in colder climates the length of the limbs decreases. In other words, according to Bergmann and Allen, people in a cold environment should be of heavy build with short limbs, and in a hot environment lean and linear. These rules relate to surface area : weight ratios – a small person, having a larger surface area : weight ratio than a larger one,[2] provides a greater surface area per unit of body mass for heat loss by convection, radiation and evaporation. Conversely, the smaller surface area : weight ratio of a large individual affords a lower surface area per unit body mass for heat loss processes. Elongation of the limbs adds to the available surface area.

While the Bergmann and Allen rules do fit certain populations in certain climatic zones (the contrasting body types of the Inuit of the Arctic and of the Nilotic people of the Sudan, for example) there are many inconsistencies, and the rules tend to be honoured more in their breach than in their observance. For instance, Strydom & Wyndham (1963) found equivalent heat tolerances in groups as morphologically diverse as Caucasoids, San, Australoids and Negroids. Indeed, Hiernaux & Froment (1976) concluded that Bergmann's rule fails to apply to sub-Saharan Africa where body size responds adaptively to the combined effects of air temperature, humidity and seasonal variation of climate.

Melanin and human thermoregulation

Melanin may contribute to the thermoregulatory systems of birds, insects, lizards, amphibians and other small animals, but its role in the human is obscure. It seems that, despite the heat-absorbing properties of melanin, there is no clear-cut relationship between melanin and heat tolerance. Negroids and other dark-skinned races do not suffer impaired heat tolerance compared to Caucasoids: in fact, they are at least equal – and, if anything, superior – in this respect. Moreover, they achieve this equality (and possible superiority) with a *lower* rate of sweating and they are therefore not compromised in terms of water balance. These findings, however, are based on hot-humid environmental conditions; the hot-dry heat of the desert may impose a different thermal strain on the Negroid, although even this is unlikely to be serious.

Thus, for people in most environments (except possibly hot-dry deserts) skin colour appears unimportant in heat tolerance. Even if melanin does absorb more heat in dark-skinned people then this has no apparent adverse effects on physiological mechanisms. Furthermore, as a radiator of heat (infrared) from the body to the environment, a deeply pigmented person has exactly the same capacity as an albino (Barnes, 1963), and so once again skin colour has no thermal advantage or disadvantage in terms of this parameter.

Concealment

Thus far we have been concerned with the functions of melanin in relation to UV and heat. But, as in the case of those bats which are darkly pigmented despite a totally sunless existence, melanin must have properties not directly associated with solar radiation. As melanin is the dominant substrate of pigmentation, it must obviously mediate all those ecological functions mediated by animal coloration. One such function is concealment.

Many animals survive only by concealing themselves from their enemies through camouflage, and the success of this camouflage must have been a potent selective force. Such animals have colours or patterns which blend so subtly with the natural background (cryptic coloration) that they escape detection from predators. The overall colour of the desert is sandy brown, and this is the colour of most of its inhabitants. In those Arctic regions where the ground is blanketed with snow and ice for much of the year, but is covered with vegetation in the short summer, animals such as the arctic hare and arctic fox moult seasonally and thereby adapt their coat colour to the changing landscape. Even the zebra and giraffe, conspicuous as they may appear in isolation, have markings which skilfully conceal them against a backdrop of trees, foliage and shadows.

Gloger's rule

Gloger, in 1833, put forward a principle which has become known as *Gloger's rule*. This states that in warm, moist climates animals and birds are dark in coloration; in dry, cold regions closely related species are pallid; and in arid desert regions they are yellow and reddish-brown. Although there are notable exceptions, Gloger's rule has been generally accepted, and it is presumed to represent a link between degree of melanin pigmentation and some kind of physiological–climatic adaptation (e.g. thermoregulation). Cowles (1959), however, interpreted Gloger's postulate on the basis of concealing coloration: in other words, animals are dark in hot, humid places not because of temperature, humidity or solar radiation but because hot, humid areas are dark as a result of the lush growth of forests and vegetation. When animals are light in hue, so is their environment. The determining criterion of an animal's pigmentation is therefore the *albedo* (i.e. the coefficient of reflection) of the terrain: the higher the albedo, the lighter the colour.

Industrial melanism

An excellent example of cryptic coloration is given by the peppered moth (*Briston betularia*). These moths make a valuable food source for numerous birds and animals and their survival depends on their camouflage ability. The typical form of the moth has a light speckled appearance and, resting on lichen-covered trunks and branches of trees, it is virtually inconspicuous. From about 1850, the industrial revolution in Europe had created widespread pollution from the combustion of coal, coke and oil; the resulting dissemination of soot had blackened trees around the cities and beyond. The peppered moth, now clearly visible against the soot-laden tree barks, was readily spotted by birds and eliminated.

A dark melanic form of the peppered moth (known as *carbonaria*) also exists. Prior to the mid-nineteenth century it was rare enough to be prized by insect collectors whereas, by the end of the century, it constituted 98 per cent of the population of peppered moths. *Carbonaria,* therefore, had almost totally replaced the typical moth, its coloration against blackened trees having made it invisible to predators. This striking evolutionary development has been called *industrial melanism,* and Kettlewell (1973), who described the phenomenon, wrote of the smoke pollution which had caused it: 'Probably no other previous cataclysm in the history of the World has produced such immediate and widespread effects upon natural history.'

There is one further act in the drama. During the mid-twentieth century smoke control measures were introduced in England, so that by 1972 atmospheric pollution had been drastically reduced. In the subsequent years not only have lichen and algae regrown on trees but the typical peppered moth of the pre-industrial era has re-emerged in all its speckled glory. Once again, natural selection has so strongly operated against the *carbonaria* that the latter is now virtually extinct in certain areas of England (Cook, Mani & Varley, 1986).

Predators

Camouflage is of biological significance not only to hunted prey but also to predators. The polar bear, whose only enemy is the human, has a white coat which reduces its visibility as it stalks seals and birds on ice floes and snow drifts.

In the African savannah, the common leopard is pale yellow with black spots. Occasionally black mutant forms arise but these quickly disappear because they are too conspicuous to be efficient hunters. But in the jungles of south-east Asia the same species of leopard is black ('black panther'), presumably because the low illumination of the rain forest favours a black predator. The same explanation applies to the gorilla (the most highly pigmented of the apes) which lives within the dense equatorial forest, whereas the more lightly-coloured baboon ranges on the savannah.

The role of camouflage in the human situation will be discussed in Chapter 11.

Animal-related functions of melanin

Warning coloration

In a sense this is the opposite of concealment. Warning (or aposematic) coloration seeks to advertise the animal and make it as conspicuous as possible. Warning colours are usually a brilliant yellow, orange or red, combined with black. Black and yellow stripes, for instance, characterize many wasps; the spotted salamander has shining yellow patches on black skin. These patterns warn would-be attackers that their potential victims will prove highly noxious or distasteful on contact – wasps with their vicious stings and the salamander with its painfully burning secretion. Predators therefore develop a conditioned avoidance reaction on sight of the warning colours. The ingenuity of nature has exploited this situation in the form of mimicry. A harmless, palatable butterfly will mimic the warning colours and markings of another butterfly notorious for its disagreeable taste; in this way the mimic deceives its enemy into leaving it well alone. Generally, it has been found that species that depend on warning coloration are much more likely to survive than those which employ camouflage.

Display

Vivid displays of colour can be effective as threats (threat display). The king cobra of tropical Asia is the largest poisonous snake and, if it is disturbed, it rears up its head to expose a brilliant orange-coloured hood: this display is usually enough to put any intruder to flight. Mammals have long black guard-hairs (usually concealed) which, by the action of special muscles, will suddenly become erect under stress to produce a threatening bristling appearance. According to Julius Caesar, the British soldiers stained themselves with a blue dye (woad) to appear more terrifying in battle.

Colour displays also have social and communication significance in the animal world, including epigamic functions (i.e. the attraction of mates during the breeding season). Some male mammals develop a brighter coat before the mating season, thus sacrificing concealing coloration for a sexual display with which to attract the female or to threaten rivals. The female, on the other hand, retains her cryptic coloration, possibly because her role in reproduction is so critical that she cannot afford to forfeit it. It is of interest that, in certain human communities, men of elevated social status (e.g. chiefs, priests, medicine-men and accomplished warriors) often resort to adornment and elaborate make-up in order to impress others with their superiority. In Western society the use of cosmetics may reflect the need to signal social and sexual desirability.

The male vervet monkey has an almost luminous blue scrotum which seems to be an important social marker and may indicate his dominance in the hierarchy. Brain (1965) observed that, when a monkey was attacked by the troop and lost his potential dominance, his scrotum changed from a brilliant blue colour to a pale powder-blue hue; at the same time he became more demoralized and less confident. This colour change is not related to the melanocyte system, but seems to be controlled by the quantity or quality of interstitial fluid above and between the dermal melanocytes (Price *et al.*, 1976).

This type of colour phenomenon in primates may have its counterpart in humans. In human males the genitalia contain one of the highest melanocyte densities in the body, while in females the melanocytes of the nipples and areolae are particularly responsive to the sex hormones. It is believed by some that pigmentation of these specific areas may represent epigamic markings and that even an occurrence such as blushing may have similar significance.

Abrasion resistance

Desert-dwelling birds are often exposed to the ravages of blowing sand. While the bird is flying its wing and tail feathers are particularly subject to abrasion from airborne particles. Thus, resistance to abrasion would constitute a potentially important adaptation to desert life. There is evidence that melanic feathers are significantly more abrasion-resistant than non-melanic feathers (Burtt & Gatz, 1982), and this may afford another explanation for the paradox that an unexpectedly large number of desert birds and desert beetles are black in colour. Indeed, selection for abrasion resistance may have been more important within the physically demanding desert environment than other selective forces such as camouflage and mimicry.

Animal colours and non-melanin pigments

Although this book is concerned with human pigmentation, it may be useful to give a brief account of the nature of animal colours. The pigmentation of humans is primitive and drab compared with that of some animals and birds. The splendour of the peacock in its panoply of colours epitomizes the complexity and extent of zoological pigmentation. The instantaneous colour change of the octopus as it glides over differently coloured backgrounds illustrates the rapidity with which the pigmentary system interacts with the environment.

Melanin is only one of several pigments contributing to animal coloration. The phaeomelanins cause the yellow and reddish colour of mammalian hair and the carotenoids are among the most widely distributed of all pigments. The latter are yellow, orange, red or violet compounds and are responsible for the conspicuous red colours of birds and fishes, for the orange colours of goldfish and ladybirds, and for the yellow feathers of the canary. Carotenoids occur abundantly in plants (e.g. in carrots); animals cannot synthesize them and they are only obtainable from the diet. Among cold-blooded species, purines and pteridines are important pigments that can be synthesized endogenously. It is the pteridines that are responsible for the bright colours of these animals.

The blue colour of animals and birds is usually not due to a pigment: it is a so-called structural colour; that is, it is due to the physical effect of light passing through a thin transparent layer. The nineteenth-century physicist John Tyndall explained the blue colour of the sky by a phenomenon now known as *Tyndall scattering*. This means that if white light passes through a medium in which very small particles are suspended some of the light strikes the particles and is scattered. The short blue wavelengths are scattered more than the longer yellow and red wavelengths, so that the medium appears blue. If the medium overlies a dark background then this background will absorb the yellow and red components and intensify the blue colour. (This optical effect accounts for the bright blue colour of human eyes.) Green is due to the Tyndall scattering of light that passes through a filter of yellow carotenoid pigment.

The melanin-containing cells (melanophores) of frog skin have the capacity to disperse melanosomes (causing skin darkening) or to aggregate them (causing skin lightening) (see p. 31). In fishes, amphibians and reptiles there are three other types of pigment cells which can translocate their granules. These are xanthophores and erythrophores (containing yellow and red pteridine-containing granules respectively) and iridophores (filled with purine-containing reflecting platelets). The latter can align their crystals in such a way as to reflect light and impart a silvery sheen or lustre to the fish or animal. The different pigment cells lie in close juxtaposition, and the intricate colour changes of these creatures (mediated both by hormones and by the sympathethic nervous system) involve the interplay and integration of all these elements. In fact, it is proposed that these fundamentally diverse cells are derived from a common stem cell which contains a primordial granule that can differentiate into melanin, pteridine or purine units (Bagnara *et al.*, 1979).

In respect of animal coloration it is important to note that colour vision is possessed by relatively few members of the animal kingdom – mainly birds and certain fish. Among mammals only humans and the higher primates are so endowed. Animals with black-and-white vision are acutely sensitive to variations in shades and tones, and in this way they achieve a high order of visual discrimination.

Notes

1 Under commercial development at present is a sunscreen cream which consists of natural human melanin incorporated in microscopic sponges.
2 With an increase in body size the surface area changes as the square but the body weight changes as the cube.

5 Non-cutaneous melanin: distribution, nature and relationship to skin melanin

This chapter deals with the occurrence of melanin in sites other than skin and hair, and it will explore the putative role of the pigment in these situations. With the exception of melanin in the eye all the other melanin-bearing tissues are internal and are totally shielded from light. Hence arises the problem of what likely use melanin would have in locations where it is deprived of its primary functions of photoprotection, thermoregulation and ecological adaptation (e.g. camouflage). Indeed, unlike beauty, melanin is not skin deep!

Eye
Iris
This is the visible pigmented region of the eye, and in describing an individual's eye colour one is referring to the pigmentation of the iris. The iris consists of several layers but, from the standpoint of colour, the two most important portions are the anterior layer together with its underlying stroma (both containing melanocytes) and the posterior pigmented epithelium.

Eye colour depends partly on the amount of melanin pigment in the anterior layer and stroma and partly on optical phenomena. In brown and dark brown irises there is an abundance of melanocytes and melanosomes in the anterior layer and stroma. Blue eyes are not the effect of a blue pigment, but represent Tyndall scattering (see p. 72). In blue-eyed individuals the anterior layer and stroma contain very little (if any) melanin. As light traverses these relatively melanin-free layers, the minute protein particles of the iris scatter the short blue wavelengths to the surface. The blue colour is heightened because the longer wavelengths (yellow and red) are absorbed by the dark background of the posterior pigmented epithelium, which also obscures the reddish hue of the adjacent blood vessels.

The intermediate shades of eye colour between light blue and black (e.g. green, hazel) result from progressive increases in the melanin content of the anterior layer and stroma, combined with decreasing amounts of scattered blue light.

Although the Negroid iris contains more abundant and denser pigmentation in the anterior layer and stroma than the Caucasoid iris, the posterior pigmented epithelium is equally pigmented in both races (Emiru, 1971).

Ciliary body and choroid

Together with the iris these two structures form the *uveal tract*. They consist of numerous layers which contain *inter alia* melanocytes and blood vessels. The colour of the interior portion of the eyeball, known to ophthalmologists as the *fundus*, depends on the quantity of melanin in the choroidal layers. The colour of the fundus varies in accordance with skin pigmentation, being orange in fair-skinned Caucasoids and a chocolate brown in Negroids – in fact, some workers find it convenient to describe the fundus as being blond, average, brunet or negroid.

Retina

The pigment epithelium of the retina is attached to the innermost layer of the choroid. Its melanocytes form a single layer of uniform hexagonal cells which are filled with large melanosomes and project fine processes between the rods and the cones. The retinal pigment epithelium has a different embryological origin from other melanocytes in the body (including those of the uveal tract): it is not derived from the neural crest but from the outer layer of the optic cup. Melanin biosynthesis by means of tyrosinase appears identical to that in the epidermis, except that experiments with chicks showed that tyrosinase activity in the retina was high at a certain stage of embryonic development but fell to zero shortly thereafter and remained so for the rest of the chick's life (Miyamoto & Fitzpatrick, 1957b).

Electromicroscopic studies of human eye pigment have confirmed that melanosome production in the retinal pigment epithelium has virtually ceased by the time of birth; whereas in the skin, hair, iris, choroid and ciliary body the generation of melanosomes continues throughout life (Feeney, Grieshaber & Hogan, 1965). This underscores a basic difference between the retinal pigment cells, which retain their melanin content for life, and the cutaneous and uveal tract melanocytes, which are in a continuous state of turnover. Furthermore, there appear to be no racial differences in the intensity of the retinal pigment epithelium whereas non-retinal eye pigmentation is correlated with skin colour.

Mucous membranes

Melanin pigmentation occurs in the oral mucosa, the presence and intensity of such pigmentation being dependent on ethnic group (Wassermann, 1974; Hedin & Larsson, 1978). In African Negroids and Australian Aborigines it is almost universal; in Negroid Americans it is common but there is marked variation in the intensity and extent of pigmentation. In these dark-skinned populations the gums are the most consistently involved (usually symmetrically), the colour ranging from light-brown to deep-brown or blue-black. The hard palate and lips are frequently pigmented, and much less often the tongue and the floor of the mouth.

In light-skinned Caucasoids pigmentation of the buccal mucosa is uncommon, occurring in about 5 per cent of subjects (Fry & Almeyda, 1968). However, although visible pigmentation is unusual, small numbers of melanosomes can be observed microscopically in the keratinocytes of the oral mucosa in up to 75 per cent of Caucasoids (Hedin & Larsson, 1978). In populations of Asian origin (e.g. Indians and Pakistanis) oral pigmentation is seen in about 40 per cent of subjects: mainly on the lips and buccal mucosa, to a much lesser extent on the gums, and hardly ever on the tongue (Fry & Almeyda, 1968).

Marked oral pigmentation occurs in several disease states where it is obviously more easily detected in Caucasoids, in whom (unlike Negroids) it is likely to be abnormal. One of the classic conditions producing abnormal oral pigmentation is Addison's disease (see p. 123 and Fig. 8.1). Another interesting link is that between gum pigmentation and cigarette smoking. Termed 'smokers' melanosis', gum pigmentation was found in 31 per cent of smokers and was attributed to a nicotine-induced disturbance of melanin biosynthesis within melanocytes (Hedin & Larsson, 1978).

Ear

In 1851 the Italian anatomist, Corti, was the first to observe pigment in the inner ear. In the human labyrinth melanin is present in several sites such as the cochlea, saccule, utricle and endolymphatic duct and sac. A very important area of melanin pigmentation is the *stria vascularis* in the cochlea (the organ of hearing). The stria vascularis is a unique epithelium because it is composed of cells from more than one embryological source: its intermediate cells consist of dendritic melanocytes derived from the neural crest (Hilding & Ginzberg, 1977; Schrott & Spoendlin, 1987) which contain melanosomes in all stages of development and resemble the melanocytes of skin and hair follicles. However, they are found in close contact with the capillary blood vessels of the inner ear, and this association with the vasculature suggests that the stria vascularis may have a function in the secretion and absorption of endolymph. There is some evidence that the amount of melanin in the inner ear is proportional to the melanin content of the iris (Wästerström, 1984). There are also well-documented genetic syndromes in several mammalian species (including the human) of hypopigmentation of the fur and skin coexisting with deafness. The occurrence of deafness in blue-eyed white cats was noted by Darwin. There is evidence (Schrott & Spoendlin, 1987) that this association of pigment anomaly and inner ear deafness is probably due to a primary developmental failure of melanocytes to migrate from the neural crest both to the skin and to the stria vascularis (where they form the intermediate cells). Melanocytes may be essential for normal strial development and for the production of the endocochlear potential (in endolymph), the presence of which is important for cochlear hair-cell function (Steel & Barkway, 1989).

Melanin is absent from the inner ear of albino animals, and experiments have demonstrated that the inner ear pigment cells of frogs respond in exactly the same way to melanocyte-stimulating hormone (i.e. by dispersion of melanosomes) and to melatonin (i.e. by aggregation of melanosomes) as do the dermal melanophores (Cherubino, 1972).

Leptomeninges

The leptomeninges are the membranes (pia mater and arachnoid mater) enveloping the brain and spinal cord. They contain dendritic melanocytes which possess a full complement of melanosomes in all stages (1 to 4) of development (Goldgeiger *et al.*, 1984). These melanocytes arise from the neural crest and their distribution is virtually confined to the meninges covering the ventrolateral surfaces of the medulla oblongata.

Is there a relationship between leptomeningeal pigmentation and skin colour? In his study of Ugandan Negroids Lewis (1969) detected a definite trend, in that light-skinned subjects had little or no pigment of the meninges while the very dark-skinned showed more widespread pigmentation. This suggests a connexion between the two melanocyte systems and it may explain the sporadic literature reports of pigmented skin naevi co-existing with leptomeningeal melanosis. Goldgeiger *et al.* (1984), however, found that melanocyte counts in the meninges did not conform to the depth of skin colour. But, as discussed in Chapter 1, the actual density of epidermal melanocytes does not vary with race; it is the extent of melanosome formation and melanization, and the rate of melanosome transfer to keratinocytes, that are the distinguishing criteria between light-skinned and dark-skinned people. Indeed, if the amounts of uveal

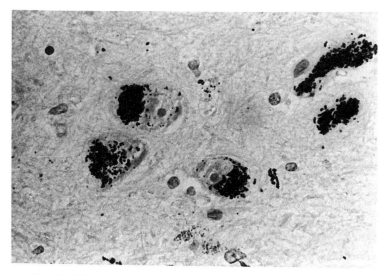

Fig. 5.1 Microscopic section of substantia nigra showing black neuromelanin granules in neurones. (Courtesy of Drs C. Warton and R. Bowen.)

tract and inner ear melanin are related to skin melanin then it seems probable that the same would apply to leptomeningeal melanin, but this aspect needs further study. The reason for the presence of melanocytes in the leptomeninges is unknown. My own speculation is that they may be involved in the regulation of cerebrospinal fluid. The latter circulates around the medulla and would therefore be in contact with leptomeningeal melanocytes. This idea is based on an analogy with the inner ear where melanocytes of the stria vascularis, being in close proximity to capillaries, appear to have a function in the secretion and absorption of endolymph.

Neuromelanin (brain pigment)

There are neurones (nerve cells) in various parts of the brain containing brown-to-black intracellular granules (Fig. 5.1); these pigment granules have been termed *neuromelanin*. The neuromelanin is distributed as a continuous column of cells extending along the brainstem but concentrated in two main areas: the *substantia nigra* in the midbrain (Fig. 5.2) and the *locus caeruleus* in the pons.

It was originally believed that pigmentation of the brain was unique to human beings. Then, when it was found also to occur in the brains of chimpanzees, orang-utans and gorillas, it was taken to be a feature of higher primates. However, more sophisticated studies have challenged the exclusiveness of neuromelanin, which is now believed to exist in almost all mammalian species. Despite this there was a kernel of truth in the old dogma, because the intensity of neuronal pigmentation is greater in primates than in other mammals and, within the primate group, its intensity increases as the relationship to humans becomes closer (Marsden, 1961).

The human brain is not endowed with its complement of neuromelanin at birth. On the contrary, it is only during the first five years of life that the pigment begins to accumulate (Fenichel & Bazelon, 1968). Pigmentation of the locus caeruleus occurs earlier than that of the substantia nigra – most cells of the former become pigmented at or shortly after birth and all are usually pigmented by 18 months whereas, at this age, the substantia nigra is still relatively unpigmented (Foley & Baxter, 1958; Mann & Yates, 1974). The amount of neuromelanin increases in a linear manner throughout childhood, adolescence and adulthood, with the pigmentation of the substantia nigra eventually equalling that of the locus caeruleus at the age of about 40 years (Mann & Yates, 1974) and exceeding it in old age (Mann & Yates, 1979). Moreover, from the age of 60 years onwards, not only does pigment deposition cease but neuromelanin concentrations actually decrease, probably because of the death of the most highly pigmented cells. It has been postulated that the excessive

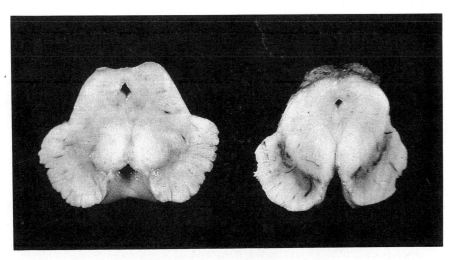

Fig. 5.2 Rostral midbrain showing pigmentation of substantia nigra in normal brain (right) and loss of pigmentation in substantia nigra of age-matched brain from patient with Parkinson's disease (left). (Courtesy of Dr Richard Hewlett.)

pigment in these cells reduces their nucleolar volume, interferes with protein synthesis and ultimately leads to cellular death (Mann & Yates, 1979). Since the substantia nigra is more heavily pigmented than the locus caeruleus in the aged it therefore bears the brunt of the degenerative processes.

Relationship between neuromelanin and cutaneous melanin

Skin, hair, eye, ear and leptomeningeal melanocytes have many features in common and they essentially enjoy membership of the same fraternity. However, there is now substantial evidence to prove that neuromelanin is in a class of its own.

The first hints of the orphan status of brain melanin emerged from autopsy studies of human albinos. Albinos lack skin, hair and eye pigment but they have normal pigmentation within the central nervous system (Foley & Baxter, 1958; Marsden, 1965). Kastin *et al.* (1976) measured neuromelanin in rats and showed that it was present in the brains of both pigmented and albino animals, whereas retinal melanin was absent in the albinos. They also administered hormones known for their direct effect on skin melanin (e.g. melanocyte-stimulating hormone) but these had no influence on the melanin of rat brain, thereby demonstrating the functional dissimilarity in the two systems.

Neuromelanin is located within neurones and not in melanocytes. For this reason malignant melanomas and naevi have never been described in the substantia nigra although such tumours have arisen within the leptomeninges (Lewis, 1969).

Since neuromelanin resides outside the body's melanocyte system, the next issue was to determine whether the enzyme tyrosinase (which is central to melanin biosynthesis in melanocytes) was involved in the production of neuromelanin. It has now been definitively shown that tyrosinase does not exist in pigmented brain cells (Barden, 1969; Rodgers & Curzon, 1975). An enzyme known as *tyrosinase hydroxylase* is prevalent in brain tissue and, like tyrosinase, it catalyses the conversion of tyrosine to dopa although, unlike tyrosinase, it cannot generate any of the subsequent compounds in the pathway leading to melanin. In spite of this tyrosinase hydroxylase is a key enzyme in brain tissue (as well as in adrenal medulla) because the conversion of tyrosine to dopa is an essential step in the formation of *catecholamines*. The brain catecholamines – chiefly dopamine and noradrenaline – function as chemical neurotransmitters, conveying nerve impulses from neurone to neurone.

Although neuromelanin has a physico-chemical profile that qualifies it as a melanin (including the possession of a stable free radical) there are definite histochemical, spectroscopic and electron-microscopic differences between it and skin melanin. Neuromelanin does, however, have properties similar to the melanins synthesized experimentally from catecholamines (Marsden, 1983). The ability of the catecholamines to be oxidized to melanin is observed clinically when black deposits of melanin (the size of pinheads) develop in the conjunctiva of the eye after instillation of adrenaline eye drops (*British Medical Journal,* 1971).

There is a striking parallelism between the location of neuromelanin in the human brainstem and the distribution of catecholamine neurones as mapped out in the brains of numerous primate and sub-primate species and in human foetal brain (Bogerts, 1981; Saper & Petito, 1982). This conspicuous overlap strongly suggests that neuromelanin is a 'waste product' of the catecholamines and that its presence is a valid marker for catecholamine neurones. This 'waste product' notion is consistent with the findings described above of a cumulative increase in neuromelanin in pigmented neurones during a person's lifetime until, in old age, these neurones are laden to the point of toxicity. The ability of the brain to form melanin from catecholamines is not dependent on enzymes (Rodgers & Curzon, 1975). Furthermore, unlike cutaneous melanin, human neuromelanin is shown by electron microscopy to contain lipid globules, an association that links neuromelanin with another brain pigment called *lipofuscin.* Lipofuscin occurs in non-pigmented neurones and it is also related to the aging process (it was once known as the 'wear and tear' pigment). The current view is that neuromelanin consists of a core of lipofuscin upon which melanin substances are deposited (Marsden, 1983).

Neuromelanin, therefore, appears to represent the accumulation of metabolic waste products in the catecholamine-containing neurones: these products are oxidized non-enzymatically to form a melanin-like compound which becomes enveloped around a nidus of lipofuscin to form the neuromelanin granule. The genesis of neuromelanin bears no similarity to the development of melanosomes within melanocytes. Its concentration in the brain is proportional to the age of the individual, and it is *totally independent* of ethnic pigmentation. Several instances have appeared in the literature where authors, erroneously assuming that amount of brain pigment corresponded to intensity of skin or eye colour, have formulated specious conclusions. The progressive pigmentation of the substantia nigra and locus caeruleus with age may partly explain the greater extent of such pigmentation in the human compared with other mammals, the former having a much longer lifespan in which to amass the catecholamine waste material. This would have confounded the early investigators who held out hopes that brain pigment signified innate human superiority – instead it seems to be a repository for neuronal garbage!

Parkinson's disease

Parkinson's disease is a neurological disorder, usually of elderly patients, which is characterized by tremor and rigidity of the limbs with slowness and reduction of movement. It was described in 1817 by James Parkinson who referred to it as the 'shaking palsy'.[1] The typical Parkinsonian patient has an expressionless face and shuffles along in short, rapid paces with a stooped posture, non-swinging arms and shaking hands. It is debilitating disorder, and one which defied understanding until the discovery in 1960 that the brains of Parkinsonian patients had markedly decreased levels of the catecholamine neurotransmitter, dopamine.

Dopamine

Dopamine occurs in the substantia nigra (the pigmented area discussed above) but its highest concentrations are in a region of brain called the *corpus striatum*. The corpus striatum has a functional relationship to the substantia nigra to which it is connected anatomically by a nerve tract. In Parkinson's disease levels of dopamine in the corpus striatum and substantia nigra are markedly reduced, and the current therapy of the disorder is to replenish dopamine concentrations by the administration of its precursor substance, dopa (in the form of L-dopa or levodopa). As noted in Chapter 2, dopa in melanocytes is converted into melanin in the Raper–Mason pathway. However, once dopa gains access to the central nervous system its metabolism follows an entirely different route. In the corpus striatum and substantia nigra it is converted into dopamine and at other sites (e.g. the locus caeruleus) it is further transformed from dopamine to noradrenaline.

Depigmentation of the substantia nigra

Loss of the normal dark pigmentation in the substantia nigra and (to a lesser extent) in the locus caeruleus is a well-established feature of Parkinson's disease and one of the most constant lesions in the disease (Figs. 5.2 and 5.3). Pakkenberg & Brody (1965) found that, although there is a decrease in all the cells of the substantia nigra in Parkinson's disease, the heaviest loss (threefold) is sustained by the pigmented neurones. A microscopic finding of great significance in Parkinsonian

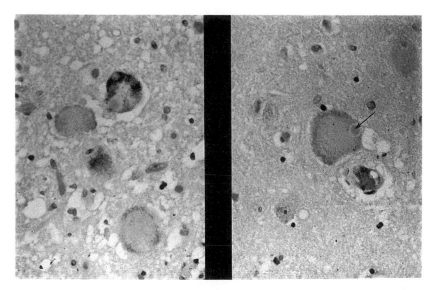

Fig. 5.3 Microscopic section of substantia nigra from brain of patient with Parkinson's disease showing (left) neurones at various stages of depigmentation (compare with normal, Fig. 5.1) and (right) well-defined Lewy body (arrowed) within depigmented neurone. (Courtesy of Dr Richard Hewlett.)

brains is the presence of eosinophilic inclusions, the so-called *Lewy bodies* (named after Lewy who described them in 1912). These are spherical, filamentous structures of undetermined nature (Fig. 5.3), and although often associated with the depigmented brainstem nuclei they also occur in non-pigmented neurones. Lewy bodies are now regarded as the pathological hallmark of Parkinson's disease (Marsden, 1983), and a clearer understanding of their nature may help to elucidate the mechanism of the Parkinsonian degenerative process.

As noted above, neuromelanin in high concentrations may produce toxicity and neuronal death. This observation, coupled with the fact that Parkinson's disease characteristically attacks pigmented neurones, would seem to indicate that the neuromelanin itself is the responsible agent in the disease. Marsden (1983) contends that, as Parkinson's disease also affects non-pigmented neurones while sparing other pigment-bearing brain areas, neuromelanin is not the primary culprit. However, a histological study on the brains of subjects who died with a diagnosis of Parkinson's disease has revealed that cell loss in the subtantia nigra and related areas directly involved the neuromelanin-pigmented neurones while the non-pigmented neurones were left relatively intact (Hirsch, Graybiel & Agid, 1988). This strongly suggests a selective vulnerability of pigmented neurones in Parkinson's disease.

Skin melanin and Parkinson's disease

Under the misconception that neuromelanin was the brain equivalent of skin melanin, Cotzias, van Woert & Schiffer (1967) injected beta-melanocyte-stimulating hormone (beta-MSH) into six Parkinsonian patients in an attempt to produce repigmentation of the substantia nigra. As expected, skin pigmentation increased but, to their surprise, the clinical signs of the Parkinson's disease became much worse! Their explanation was that the stimulation of skin melanogenesis by beta-MSH would have diverted dopa from brain to skin, thereby further depleting the deficient brain dopamine levels.

In the belief that degeneration of the pigmented neurones in Parkinson's disease might have its counterpart in skin melanocytes, Parkes *et al.* (1972) examined the skin of 102 Caucasoid patients with Parkinson's disease for any signs of pigmentary anomalies. There was no evidence whatsoever for the involvement or destruction of epidermal melanocytes in this disease.

Race and Parkinson's disease

While there appears to be no direct relationship between the loss of neuromelanin in Parkinson's disease and skin pigmentation the matter has often been raised of whether peoples of varying skin pigmentation have different susceptibilities to the disease. Surveys (both community-based and hospital-based) in Baltimore, United States of America, found that blacks had a significantly lower frequency of Parkinson's disease than did whites (Kessler, 1978), although a door-to-door survey of a biracial rural county in Mississippi revealed no substantial differences in the age-adjusted prevalence ratios among blacks and whites (Schoenberg, 1986). A study of South African hospital records showed that Parkinson's disease in blacks was far less common than in either whites or Indians (Asians) (Cosnett & Bill, 1988), and an investigation on the island of Sardinia disclosed that Sardinians (who are of mixed South European and North African descent) had a prevalence rate for Parkinson's disease half that of North Europeans (Rosati *et al.*, 1980). Thus there is some epidemiological evidence that dark skin colour may protect against the emergence of Parkinson's disease (Lerner & Goldman, 1987).

From the foregoing comments it is clear that this protection is not based on any relationship with the pigmentation of the substantia nigra. Moreover, albinos do not appear to have an increased risk of Parkinson's disease. Theoretically, as the beta-MSH experiment of Cotzias *et al.* (1967) demonstrated, the opposite scenario would be predicted: so much of the body's dopa is mortgaged to skin melanogenesis in dark-skinned people that there should be insufficient for the brain and thus an increased susceptibility to the disorder. But Kessler (1978) attributed the lower risk of Parkinson's disease in blacks to skin melanin, and he speculated that pigmented tissues may be more resistant to invasion by neutropic viruses. This emphasis on an environmental agent (not necessarily viral) has been confirmed by recent concepts of the disease (see below). Heredity itself appears relatively unimportant in the transmission of Parkinson's disease and twin studies have failed to provide evidence of a significant genetic component (Snyder & D'Amato, 1986).

MPTP and Parkinsonism

In 1977 a 23-year-old college student developed an acute state of Parkinsonism[1] which, after meticulous investigation, proved to be due to the compound methylphenyltetrahydropyridine (known by the abbreviated form *MPTP*]. The patient was a drug addict and he had set up a home-based laboratory for the illicit synthesis of drugs. During the procedure he had inadvertently produced MPTP, an analogue of pethidine, and, after administering this intravenously, had succumbed to severe Parkinsonism. Parkinsonism has also occurred in other young people after the self-administration of similarly synthesized preparations. MPTP is a neurotoxin which selectively destroys the pigmented cells of the substantia nigra. Humans and non-human primates are highly sensitive to MPTP neurotoxicity, whereas rats and mice are insensitive. This observation elicited the idea that, because the former have abundant neuromelanin and rats and mice have a sparse distribution, the toxicity of MPTP may be mediated by neuromelanin. However, it was demonstrated that the toxicity does not reside in MPTP itself but in its active metabolite, the methylphenylpyridine ion (MPP$^+$), which has a powerful affinity for melanin. On the basis of this finding, the following hypothesis was derived (Snyder & D'Amato, 1986): MPTP is concentrated in the substantia nigra where it attaches to an enzyme (monoamine oxidase) that metabolizes it to MPP$^+$. MPP$^+$ is specifically taken up into the neurones of the substantia nigra where it is avidly bound to neuromelanin and from which it is then gradually released into the cell in concentrations sufficiently toxic to cause neuronal damage and death. This bizarre discovery of MPTP has triggered a new and vigorous research thrust into Parkinson's disease and, as with so many other outstanding discoveries in medicine, the first prize must go to serendipity.

MPTP has given renewed impetus to the search for environmental toxins in the cause of Parkinson's disease. The contribution of industrial chemicals, herbicides and pesticides (some, e.g. paraquat, structurally similar to MPTP and MPP$^+$) to the development of Parkinson's disease has been emphasized by certain workers (Bocchetta & Corsini, 1986; Chapman *et al.*, 1987) and challenged by others (Vieregge, Kömpf & Fassl, 1988). Manganese is another ion that can induce Parkinsonism; it has a high affinity for melanin-containing tissues and it appears to damage neuromelanin-bearing neurones (Lindquist, Larsson & Lyden-Sokolowski, 1987). If the assumption is made that Parkinson's disease may result from the binding of an environmental toxin to neuromelanin, then the apparently decreased prevalence of the disease in dark-skinned groups may be due to a greater binding of potential neurotoxins to skin melanin with a consequent lowering of their access into brain tissue (Lerner & Goldman, 1978).

Cigarette smoking and Parkinson's disease

Cigarettes are rightly denounced as a health hazard. But an unexpected trend has been emerging – cigarette smoking appears to have a *protective* effect on the development of Parkinson's disease. There is a negative correlation between cigarette smoking and the incidence of Parkinson's disease such that smokers have a 20–70 per cent reduction in risk of the disease (Barron, 1986). There is no explanation for this although several theories have been advanced. In short, the thinking is that cigarette smoking influences dopamine function in the substantia nigra and corpus striatum. Nicotine may be implicated, but it is more probable that the carbon monoxide in cigarette smoke may protect the substantia nigra by trapping the noxious free radicals generated in metabolic reactions. Another theory (Boulton, Yu & Tipton, 1988) is that, as cigarette smoke inhibits monoamine oxidase activity, it will prevent the conversion of MPTP to MPP$^+$ and thereby protect against the resultant neurotoxicity.[2] However, the validity of this relationship between smoking and Parkinson's disease requires confirmation, and writers have been quick to discourage any attempts to introduce smoking as a preventive treatment. It is highly unlikely that the overwhelming legacy of harm and damage from cigarette smoking can be offset.

Notes

1 The accepted terminology is to refer to the primary, naturally occurring disorder (as described originally by James Parkinson) as 'Parkinson's disease' and to the neurological disorders resembling it, but secondary to other factors or causes (e.g. drugs, toxins, viruses), as 'Parkinsonism'.

2 There is evidence that a monoamine oxidase (type B) inhibitor, selegiline (deprenyl), may slow the rate of neuronal degeneration in Parkinson's disease.

6   The properties and possible functions of non-cutaneous melanin

The circulation of melanin

The melanosomes in epidermal melanocytes are transferred to keratino-cytes, thereafter conveyed upwards to the stratum corneum, and eventually disposed of by desquamation. There is also a less well-known transaction that proceeds in the opposite direction: a circulation of melanin from the skin to the internal organs of the body.

The knowledge that melanin is manufactured in the skin by *in situ* pigment cells dates from relatively recent times. The early nineteenth-century notion was that black pigment originated in the bile, and ingenious theories were devised of how bile turned itself into the colouring material of the skin. In an informative review of the literature, Wassermann (1965a) showed that the circulation of pigment was actually the first aspect to be described. A hundred years ago it was thought that wandering cells, probably leucocytes (white blood corpuscles), sca-venged red blood cells and carried them and their pigment (haemoglobin) to the epidermis for its nutrition, the haemoglobin then being trans-formed into melanin. Even though these ideas may now seem preposter-ous, the concept of a circulation of melanin has recently been revived and revised.

The dermis contains an extensive network of lymphatic ducts which drain into lymph nodes. There is a much higher incidence of melanin deposition in the skin-draining lymph nodes of Negroids than of Cauca-soids (Wassermann, 1965a). This suggests that melanin is transported by lymphatic drainage into the circulation from which it can then be distributed throughout the body.

The modern approach to the circulation of melanin stems chiefly from the work of Wassermann (1974). He used the skin-window technique to show the inflammatory response in the skin of South African Negroid and 'Cape Coloured' subjects. In simple terms, he found that leucocytes in the inflamed area actually engulfed melanosomes in the same way that they did foreign matter, and that these melanin-tagged leucocytes then re-entered the circulation via lymphatic channels. (This phenomenon was seldom observed in the Caucasoid skin response.) Pigmented leucocytes

88

have been identified in the peripheral blood of Negroids and 'Cape Coloureds', but not in that of Caucasoids. Wassermann maintained that the leucocytic transport of melanin fulfils an overflow function in situations where the epidermal melanin unit is over-burdened (e.g. in very dark skin or in disorders of hyperpigmentation). Furthermore, because melanin can neutralize free radicals, the transport system can convey it to metabolically active tissues with high free radicals concentrations.

Drug binding to melanin

The previous chapter described the stable free radical property of melanin, a property which enables it to trap noxious free radicals such as those generated in cells by UV. Melanin is a strong electron acceptor and this makes it a potential binding site for compounds such as MPP^+. There are several medicinal agents in common therapeutic use which bind to melanin and thereby cause clinically undesirable side-effects.

Phenothiazines

These drugs are primarily used as antipsychotic agents and specifically in the treatment and management of schizophrenia. The prototype of the phenothiazine group of drugs is *chlorpromazine*.

Before the phenothiazines were introduced in the early 1950s the prognosis of schizophrenia was poor and many patients remained in mental hospitals for life. The advent of the phenothiazines revolutionized the outcome of the illness. These drugs counteract the disturbing psychotic features of schizophrenia (such as the delusions and hallucinations), make patients more manageable and often allow their discharge from hospital. Unfortunately the phenothiazines are not curative, and a substantial number of schizophrenics will relapse if treatment is discontinued. Long-acting phenothiazine injections are available and these are effective for up to a month, so that they provide a reliable means of long-term administration.

The action of the phenothiazines is to block dopamine receptors in the brain with the resultant development of Parkinsonism (*drug-induced Parkinsonism*) (see note 1, Chapter 5). The latter is reversible on stopping the drug or reducing the dosage but, if necessary, there are other agents (anticholinergic drugs – and not dopa) that readily treat and prevent the problem. There is another condition, known as *tardive dyskinesia,* which arises after protracted use of phenothiazines; it takes the form of involuntary movements, usually of the mouth, lips and tongue but also occasionally of the limbs and trunk. It bears no resemblance to drug-induced Parkinsonism and it is generally unresponsive to any kind of treatment. Tardive dyskinesia may reflect some degree of brain damage: elderly patients and those with organic brain disease are most susceptible to it.

It is of great interest that there is a link between the phenothiazines and melanin pigmentation. Potts (1962) showed that radioactive phenothiazines are specifically taken up by the melanin-bearing uveal tract of the eye (except in albino animals, which lack melanin). Lindquist (1973) found that radioactive chlorpromazine rapidly crosses the placenta and accumulates in the eye and inner ear of the pigmented (but not albino) foetus. Furthermore, when human brain sections were incubated in radiolabelled chlorpromazine the most intense radioactivity appeared in the neuromelanin-containing neurones of the substantia nigra and locus caeruleus. This selective affinity of chlorpromazine for pigmented tissues is due to the physico-chemical properties of melanin. Potts (1964) isolated pigment granules from the uveal tract and retinal pigment epithelium and mixed these with chlorpromazine: very high concentrations of the latter were bound by the granules. Chlorpromazine is an excellent electron donor, melanin an equally good electron acceptor, and chlorpromazine was thus able to form a strong bond with melanin as a result of an electron exchange called a *charge transfer reaction* (Potts, 1964).

The powerful adsorption of phenothiazines on to melanin leads one to ask whether these drugs are harmful to schizophrenic patients. Their localization in the eye may be responsible for the pigmentary retinopathy caused by a particular phenothiazine (thioridazine) in high dosages. Does the binding of phenothiazines to the substantia nigra influence the emergence of drug-induced Parkinsonism? The latter is far more likely to be due to the blockade of dopamine receptors, but the affinity of these agents for neuromelanin may have as yet undetermined consequences. Phenothiazine-induced tardive dyskinesia has been associated with degerative brain changes, including those to the substantia nigra (Lohr, Wisniewski & Jeste, 1986). A recent case report (Ball & Caroff, 1986) documented the co-existence of pigmentary retinopathy and tardive dyskinesia in a patient receiving thioridazine – both conditions, it was speculated, being attributed to the affinity of the phenothiazine for melanin (in retinal pigment and neuromelanin respectively). Research into these aspects has been sparse and the problem of drug binding to melanin (with its potential toxicity) warrants much more investigation.

The clinical syndrome of the skin–eye pigmentation induced by phenothiazines will be discussed in Chapter 8. This syndrome may have some bearing on the phenothiazine–melanin interaction.

Chloroquine

Chloroquine is best known as an antimalarial drug although it is used in various other medical conditions such as rheumatoid arthritis. Like the phenothiazines chloroquine accumulates selectively in the pigmented tissues of the eye and the inner ear (Lindquist, 1973). It has a marked affinity for melanin and its toxicity appears to be greater than that of the phenothiazines, especially if it is prescribed for non-malarial conditions which require high dosages for long periods. There have been numerous reports of *chloroquine retinopathy,* a serious disorder which leads to irreversible loss of vision and which can occasionally manifest years after the cessation of treatment. The primary lesion is in the retinal pigment epithelium, the cells of which enlarge and increase their content of melanosomes. Clumps of pigment migrate into the layers of the retina with destruction of the sensory nerve cells (rods and cones). Animal experiments have shown that albinos are immune to the retinal toxicity of chloroquine – this appears to be one of the very few advantages that albinos have over their pigmented fellows!

The predilection of chloroquine and the related antimalarial compound, quinine, for inner ear melanin (particularly in the stria vascularis) may explain their toxic effects on hearing (ototoxicity). Loss of hearing has occasionally followed chloroquine administration, and one alarming report described the occurrence of deafness in two children born to a mother who had received high dosages of chloroquine during her pregnancies (Hart & Naunton, 1964). Ototoxicity is a well-known feature of a group of antibiotic drugs known as the aminoglycosides (they include streptomycin and kanamycin) and, like chloroquine, these compounds have a high affinity for melanin (Lindquist, 1973).

There have been reports of acute neurological reactions following chloroquine therapy for malaria (Singhi, Singhi & Singh, 1979; Khilnani *et al.*, 1979). These include stiffness of the neck, jaw and limbs, protrusion of the tongue and fixation of the eyeballs, features that are closely related to drug-induced Parkinsonism and indicate involvement of the substantia nigra and corpus striatum. It is therefore possible that the affinity of chloroquine for neuromelanin may be responsible for these reactions. It is of interest, too, that this particular property of chloroquine has made it a potent competitor for MPP^+ binding to neuromelanin (see p. 85), and chloroquine administration to monkeys before they received MPTP protected the animals from MTPT-induced Parkinsonism and substantia nigra damage (D'Amato *et al.*, 1987).

Atropine

The influence of melanin on a drug's action is well illustrated in the case of atropine and the eye. Atropine is an anticholinergic agent that causes enlargement of the pupil. Pupillary dilatation is often required by ophthalmologists to enable them to make a proper examination of the interior of the eye, and for this purpose atropine eye drops are instilled. It is a common clinical observation that the pupil in Negroids dilates more slowly and to a smaller extent in response to atropine drops than the Caucasoid pupil (Emiru, 1971). The probable explanation for this difference is the deeper pigmentation of the Negroid iris. The latter would increase the binding of atropine to melanin and cause a slower release of the drug on to its specific receptor sites. This difference is not a racial one because albino Negroids showed the same briskness of pupillary response to atropine drops and the same degree of dilatation as Caucasoids (Emiru, 1971).

The same principle applies to the local anaesthetic agent, lidocaine, which is reversibly bound to the melanin of the inner ear. Lidocaine has a place in the treatment of disabling tinnitus (noises in the ear) and the hypothesis is that its accumulation in inner ear melanin creates a storage depot from which the drug can be released to exert its local anaesthetic action (Wästerström, 1984).

Ethnic differences in drug response

A rational enquiry at this stage would be whether different ethnic groups respond differently to medicines which are specifically adsorbed onto melanin. If melanin is a strong binder of such drugs then theoretically a very pigmented person would sequester more of the drug in the skin and hair than a fair-skinned individual, and this should reduce the drug's availability and therefore both it efficacy and its adverse effects.

Literature on this subject is limited. There is some evidence that, compared with Caucasoids, Mongoloid patients require lower therapeutic dosages (weight-standardized) of antipsychotic drugs for optimal clinical response and have a greater susceptibility to drug-induced Parkinsonism (Binder & Levy, 1981; Lin *et al.*, 1989). However, these findings have not been corroborated by other studies (Sramek, Sayles & Simpson, 1986; Binder *et al.*, 1987). A hospital inpatient survey showed similar rates of drug-induced Parkinsonism in South African Negroids, Asians (Indians) and Caucasoids (Cosnett & Bill, 1988). Furthermore,

Caucasoids have a prevalence of tardive dyskinesia which is comparable with that of South African Negroids (Holden, 1987), Nigerian Negroids (Gureje, 1987) and Japanese (Binder *et al.*, 1987). Therefore, in sum, the scanty data currently available do not support the hypothesis that dark-skinned patients are *less* susceptible to either the therapeutic or the adverse effects of melanin-binding drugs.

The possible functions of non-cutaneous melanin

In Chapter 4 the role of integumentary melanin was discussed and several probable functions formulated from the existing body of empirical evidence. However, the functions of the non-cutaneous melanins remain obscure, and even speculations are tenuously based owing to the lack of an established corpus of knowledge. It is also only in recent times, with the renewed interest in Parkinsonism, that the focus has been on neuromelanin – although pigmentation of the substantia nigra was commented on by Vicq d'Azyr in 1786. Whereas the biology of skin melanocytes has been frenetically investigated over the past four decades pigment cells elsewhere have been overlooked, partly because of their inaccessibility. In addition there has been the intuitive notion that melanin occurring in the non-illuminated recesses of the body cannot have any credibility. Perhaps recent disclosures about the central position of melanin in mediating drug-induced neurotoxicity, retinal toxicity and ototoxicity may introduce a different dimension to the melanin research of the future.

Ocular melanin

A priori it would seem that ocular melanin limits the scattering of light rays within the interior of the eye and thereby reduces the amount of light impinging on the sensory nerve cells (photoreceptors) of the retina. This shielding action would maintain visual acuity under brightly lit conditions. Albinos, who lack eye melanin, have severe visual handicaps: their hypersensitivity to the sun makes them photophobic and they cannot venture into the daylight without dark glasses – hence their designation as 'moon children'.

The albino is an extreme example which has actually led anthropologists into misconstruing the function of melanin in dark-eyed and light-eyed individuals respectively. Extrapolation from albinism suggests that the former would have greater visual acuity in bright light and be better adapted to stressful environments such as deserts and snowfields. But two independent studies (Hoffman, 1975; Short, 1975) have scotched this idea – the density of eye pigmentation was found not to influence visual acuity under increasing conditions of brightness. These negative reports were a blow to evolutionists who had convincingly located populations with the darkest eye colour in geographical regions with the strongest illumination.

The advent of light-coloured eyes in Western Europe was also slotted into an evolutionary niche. At first it was believed that light-eyed individuals perceived shorter wavelength colours (i.e. violet, blue) more acutely than did their darked-eyed fellows. This supposed attribute was seized upon as being advantageous to the Pleistocene hunters of glacial Europe who could therefore identify their prey in misty and foggy weather. Dodt, Copenhaver & Gunkel (1959) quashed this theory when they found no difference in the sensitivity of Caucasoid or Negroid eyes at the blue end of the spectrum, although they did show that Caucasoids were more sensitive to red light than Negroids, and albinos had the greatest sensitivity of all. Accordingly, light eyes were deemed useful to the ancient cavemen of Europe who needed to see and hunt cave bears by firelight! This *ad hoc* theorizing, ingenious as it may be, reflects a basic uncertainty as to the adaptive function of ocular pigmentation.

Iris colour and psychomotor performance

Worthy (1974) put forward an interesting hypothesis. He claimed that dark-eyed individuals performed better at 'reactive activities' whereas light-eyed people were superior at 'self-paced activities'. A reactive activity is one where a response must be made appropriately and timeously to a situation over which the individual has little control. Examples of sports activities that are largely reactive in nature are boxing and basketball. A self-paced activity, on the other hand, requires an individual to choose his or her own time in which to respond to a situation. Golf and bowls are examples of self-paced sports.

Worthy & Markle (1970) studied the statistics for various sports and concluded that Negroids were better at reactive than at self-paced games. There were few Negroid golfers or bowlers but excellent Negroid boxers. Even within a sport Negroids were better on the reactive tasks – for example, in baseball they were significantly more likely to be non-pitchers (reactive) than pitchers (self-paced); in basketball, Caucasoids were inferior to Negroids in field goal accuracy (reactive) but superior in free-throw accuracy (self-paced). In professional bowls light-eyed bowlers had higher money winnings than their dark-eyed counterparts! In the animal kingdom, too, a similar relationship prevails. Animals and birds that hunt by the self-paced techniques of stalking and surprise (e.g. cats, herons) are light-eyed whereas those that capture prey in a reactive manner by taking birds or insects 'on the wing' (e.g. swifts, swallows) are dark-eyed.

Attempts have been made to corroborate Worthy's hypothesis and with some success. Several groups have demonstrated that dark-eyed subjects had faster reaction times than light-eyed subjects (Landers, Obermeier & Patterson, 1976; Tedford, Hill & Hensley, 1978; Hale *et al.*, 1980). The faster reaction times were in response not only to visual but also to auditory stimuli. None of the investigating teams was able to offer a satisfactory explanation for their findings.

Inner ear pigmentation

Various studies have examined the association between inner ear melanin and auditory function, and these have been briefly reviewed by Wästerström (1984) and Barrenäs & Lindgren (1990). On the basis of an observation that the amount of melanin in the inner ear was proportional to its concentration in the iris, a number of workers have assessed the hearing capacities of blue-eyed as compared with brown-eyed subjects. Although there are some conflicting results the general outcome is that blue-eyed individuals suffer more auditory fatigue and have poorer hearing than do those with brown eyes. Furthermore, when possible ethnic differences have been sought, several surveys attested to superior hearing in Negroids compared with Caucasoids.

It therefore seems that the depth of inner ear pigmentation (as judged by the depth of iris pigmentation) tends to be positively correlated with auditory function. Albino guinea-pigs display lower auditory thresholds than pigmented animals and are more susceptible to the traumatic effects of high intensity noise (Conlee *et al.*, 1986a) while human albinos also show evidence of auditory abnormalities (Garber *et al.*, 1982). An experimental study of male Caucasoid (Swedish) teenagers found that those with darker complexions are more protected against noise-induced temporary hearing loss than their less pigmented (and more sun-sensitive) fellows (Barrenäs & Lindgren, 1990).

A possible role of non-cutaneous melanin

It appears that in normal (i.e. non-albino) human beings an increase in iris and inner ear pigmentation is associated with superiority in specific aspects of neurological function. It is possible that these differences in function may be related to the actual presence of melanin within the tissues.

Skin melanin has an obvious role in photoprotection. Its stable free radical property enables it to capture harmful free radicals liberated in the skin by UV radiation. There is much to suggest, too, that this free radical component of melanin underlies its functions elsewhere in the body.

McGinness, Corry & Proctor (1974) demonstrated that melanin can act as an amorphous semiconductor threshold switch, that is, it can convert and transfer energy generated in a biological system. These physical attributes of melanin are complex but, put simplistically, they neutralize high energy states in the cell by converting them into less harmful modes (e.g. the absorbed UV radiation is transformed into heat). Sound, like light, can also produce high energy states, and it seems logical that the melanin in the inner ear (cochlea) would dissipate this energy and protect the organ from acoustic damage. Conversely, just as a deficiency of skin melanin predisposes to sunburn and eventually to skin cancer, so deficient cochlear melanin adversely affects hearing, especially if the noise input is excessive (Conlee *et al.*, 1986a). Furthermore, the melanocytes in the stria vascularis of the chochlea seem to be implicated in the production and maintenance of endolymph (see p. 76) and cochlear melanin may thus exert a protective function at the microvascular level.

Drugs such as chloroquine and the aminoglycoside antibiotics bind strongly to melanin and produce toxic effects in the eye and ear. It is possible that, by forming a complex with melanin, these drugs inactivate the pigment and thereby render the tissue more vulnerable to high energy states and to the resultant damage. This concept may have some value in the treatment of melanoma. For example, the addition of melanin-binding drugs (such as chlorpromazine or kanamycin) to melanized cells increases the killing efficiency of ultrasound radiation on these cells (Wästerström, 1984). Such agents, therefore, should achieve a dual purpose if administered to patients with melanomas: they would concentrate selectively in the melanoma tissue and then sensitize such tissue to radiation therapy.

Neuromelanin in the pigmented brain areas is probably not an inert substance. Neuromelanin will store energy dependent on the ionic and electrical gradients produced by the neurone (McGinness, 1985). In neuromelanin-containing neurones the action potential system and the neuromelanin may interact through phonon–electron coupling (Lacy, 1984). Through these interactions neuromelanin may influence the firing frequency of neurones and this may allow pigmented neurones to process information differently from their non-pigmented counterparts. Drugs which bind to neuromelanin (e.g. phenothiazines) are likely to modulate the electronic properties of the pigment.

A dynamic and contentious essay by Barr (1983) singles out melanin as the fundamental organizational molecule in living systems because its properties enable it to regulate and integrate the molecular, metabolic and endocrine networks within the body. With regard to neuromelanin, Barr maintains that its strategic situation along the brainstem – the gateway to and from the higher centres – makes it the physical substrate for consciousness, the emotions and indeed the 'mind'. Although his claims for neuromelanin are exaggerated and even grandiose, Barr does succeed in highlighting the versatile potentials of a substance that has hitherto been regarded as a waste produce of catecholamine metabolism.

7　*Measurement of skin colour*

Measurement techniques

The scientific study of human skin colour requires accurate measuring instruments. Before the latter became available verbal descriptions of skin colour had to suffice. It is true that terms such as 'black', 'brown' and 'white' (or 'blond' and 'brunet') are universally understood and each can be further qualified as light, medium or dark, but verbal descriptions such as these are only really applicable to broad categories and they are not discriminating enough to identify *gradations* of skin colour within and between populations. In the same way, a verbal description of a green paint would be totally inadequate for the selection of an accurate colour match. Furthermore, verbal reporting relies on subjectivity and this often introduces bias and distortion. Any form of measurement which seeks scientific acceptability must use standardized methods that are not only reliable and objective but easily reproducible.[1]

In order to meet these needs various colour-matching techniques have been devised (e.g. Gates's tinted papers). The most widely used of these has been von Luschan's skin colour tablets. These consist of a set of small ceramic tiles consecutively numbered from 1 to 36 and ranging from pure white to black. There have also been colour atlases (e.g. the Munsell system and the Medical Colour Standard for Skin) which have provided sets of standard colour chips for skin matching. However, these methods have the disadvantages not only of subjectivity but that the surface texture of the chip or tile may not resemble that of the skin (or may have deteriorated with time) so that matching becomes imperfect.

Another colour-matching method was the Bradley colour top. This comprised four discs – white, black, yellow and red – which, when spun, produced a colour dependent on the proportionate blending of each disc. (The matching colour could be designated by the percentage admixture of each coloured disc, although in practice the black component was usually taken as the index.) This procedure permitted a quasi-numerical measure of skin colour but it also depended too heavily on the subjectivity of visual matching. The colour top is now obsolete, but it did offer early investigators a means for research and it was used in the pioneer studies of skin colour inheritance.

Reflectance spectrophotometry

The modern era in skin colour measurement began in 1926 with the use of the recording spectrophotometer. This instrument was modified and improved by Hardy in the 1930s, and Edwards & Duntley (1939b) applied it to their now classic analysis of the colour properties of living human skin. The Hardy recording spectrophotometer successively illuminates the skin area through all the visible wavelengths (400–700 nm) and then automatically plots a curve of the percentage of light reflected from the skin (skin reflectance) over all of these wavelengths. Thus 'skin colour' is measured not as a visual parameter but in terms of its reflective properties. This eliminates the fallibility of human observation and judgement and brings the assessment of skin colour within the scope of accurate and objective measurement.

The Hardy apparatus was cumbersome and unsuited to field-work, but it was not until the early 1950s that the portable 'EEL Reflectance Spectrophotometer' came into being. This machine was manufactured by Evans Electroselenium Ltd, Harlow, Essex, and it was Weiner (1951) who first described its usefulness in skin colour measurement. It was primarily invented for the paint and dye industries but its portability and simplicity made it ideal for field surveys. Indeed, from the time of its launching right up to the present day the EEL machine has held sway amongst anthropologists as the most popular barometer of skin colour. Its extensive use throughout the world is apparent from the studies tabulated in Tables 7.1–7.4.

The EEL reflectance spectrophotometer (Fig. 7.1) is capable of giving objective and reproducible measurements of pigmentary variation on a continuous scale of reflectance. It is designed in the form of two main units – a galvanometer unit and an applicator head – and it is powered by either mains or battery. The applicator head is freely mobile and can be applied to the skin surface in most regions of the body. It contains a 6 volt, 6 watt tungsten lamp, the light from which is focused to form a circular spot on the skin. The colour of the light is modified by intercepting different filters: the latter are mounted on a wheel and they are rapidly interchangeable by a flick of the thumb. The light strikes the skin at an angle of 45 degrees; it is reflected by the skin to fall on a photocell located in the applicator head. The photocell then generates a current which is transmitted to the galvanometer unit and is measured there. The reading represents the percentage reflectance of light relative to that from a standard. The latter is a pure white magnesium carbonate block to which the instrument is set at a reflectance of 100 per cent. Readings are obtained for each of the nine filters on a linear scale calibrated from 0 to

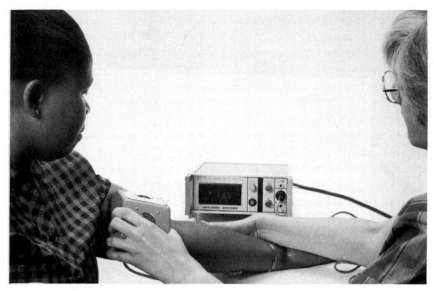

Fig. 7.1 Application of EEL reflectance spectrophotometer. Reading of 13.5 represents percentage reflectance of light at Filter 601 (425 nm) from skin of inner (medial) aspect of upper arm. (Courtesy of Dr Jennifer Kromberg.)

100, and the later models of the instrument (Fig. 7.1) also provide the option of a digital display readout unit (accurate to 0.1 per cent) instead of the mirror galvanometer (accurate to 1 per cent). These values represent the percentage of light reflected by the skin at each filter: the lighter the skin colour, the higher is the reflectance. The nine filters (referred to as Ilford filters 601–609) 'sample' the whole of the visual spectrum from the violet end (filter 601) to the red (filter 609). Therefore, unlike the Hardy apparatus, the EEL instrument gives an abbreviated coverage of the visual spectrum. The nine filters transmit with the following dominant wavelengths:

> 601–425 nm (violet); 602–465 nm (blue); 603–485 nm (blue-green); 604–515 nm (green); 605–545 nm (yellow-green); 606–575 nm (yellow); 607–595 nm (orange); 608–655 nm (red); 609–685 nm (deep red).

It is customary to plot the percentage skin reflectances (ordinate) against the dominant wavelengths of the nine filters (abscissa). This produces a spectral curve which gives a quick graphic representation of the 'skin colour' profile of an individual (Fig. 7.2). Such spectral curves are particularly useful in highlighting the pigmentary differences between population groups.

Guidelines are available for the practical use of the EEL machine during the testing procedure (Weiner & Lourie, 1981). Almost all studies have taken reflectance readings from the inner (medial) aspect of the upper arm, approximately mid-way between the axilla and the medial epicondyle of the humerus. This area is relatively unexposed to the sun and it is a convenient site because the rim of the applicator head can be placed against the medial epicondyle of the humerus. Although this is not generally mentioned, the position of the upper limb should be standardized during the recording of reflectances because postural changes in the limb affect readings taken from the same site (Buckley & Grum, 1961). The usual practice is to test with the upper limb either abducted to 90° (i.e. in a coronal plane) or flexed to 90° (i.e. in a sagittal plane). If used correctly, reflectance spectrophotometry is free of any significant inter-observer error (Lees, Byard & Relethford, 1978). Furthermore, under normal conditions of field survey it is not important to control for environmental temperature (Little & Sprangel, 1980).

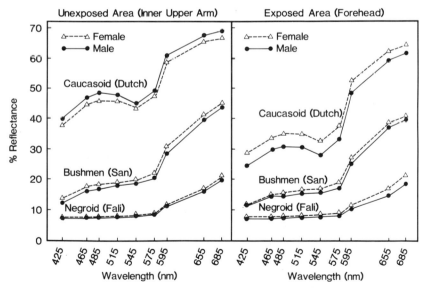

Fig. 7.2 Graph of skin reflectances taken at nine wavelengths of EEL reflectance spectrophotometer. Measurements obtained from unexposed area (inner upper arm) and exposed area (forehead) in males and females of three population groups differing markedly in skin colour. (Data sources – Dutch: Rigters-Aris, 1973b; San: Tobias, 1961; Fali Tinguelin: Rigters-Aris, 1973a.)

Photovolt reflectance spectrophotometer

The EEL instrument is not the only portable one for measuring skin pigmentation. There is the Photovolt machine, manufactured by Photovolt Corporation, New York. Both reflectometers operate on essentially similar principles, the major difference between them being that they sample the visible spectrum at different wavelengths. They contain different coloured filters whose dominant transmission characteristics vary across the spectrum (Lees & Byard, 1978). The Photovolt spectrophotometer uses three main filters (blue, green and red) and three tristimulus filters (triblue, trigreen and triamber) which transmit with the following dominant wavelengths (Byard, 1981):

> blue – 420 nm; triblue – 450 nm; green – 525 nm; trigreen – 550 nm; triamber – 600 nm; red – 670 nm.

The advantages of the Photovolt machine are that it is both easy to use and much more robust as a field instrument than the EEL. Its disadvantage is the wide band transmittance of the filters: this property reduces the precision of measurement (specificity) of each filter. (The EEL reflectometer has narrow waveband filters.) Another difficulty is that the great majority of skin reflectance studies in the world are based on EEL measurements; however, several important studies emanating from the Americas have employed the Photovolt equipment. Unfortunately the dissimilarity of the two instruments vitiates any basis of comparability between their respective data.

Garrard, Harrison & Owen (1967), however, attempted to establish comparability by deriving linear and quadratic regression equations to achieve conversions from one machine to the other. These equations were not entirely adequate (Lees & Byard, 1978) and the use of multiple-regression techniques has produced more reliable and accurate conversion coefficients, although separate formulae are still required for dark-skinned (Lees & Byard, 1978) and light-skinned populations (Lees, Byard & Relethford, 1979). For any given wavelength, the Photovolt reflectances are lower than the corresponding EEL values, and this applies especially to the light-skinned groups.

Fibre optics

One of the major practical limitations of the above-mentioned spectrophotometers is the difficulty in positioning the reflectance head correctly on certain areas, such as the face, without causing deformation of the skin surface and resultant colour changes from the altered vascularity (see below). This problem has been overcome by the use of flexible fibre optic light guides. Smooth filaments of transparent materials, such as glass, conduct light with high efficiency, and a practical light guide consists of a bundle of several thousand fibres cemented together at their ends. The basic applications of fibre optics to spectrophotometry are discussed by Gibson (1971).

Modern and sophisticated reflectance spectrophotometers now incorporate optical fibres and computer technology (Bjerring & Andersen, 1987). These allow for the rapid and accurate monitoring of skin colour changes (due to disease or vascular abnormalities) which may be useful to dermatologists.

Pigments contributing to skin colour

The work of Edwards & Duntley (1939b) with the Hardy recording spectrophotometer showed that five pigments were responsible for the colour of normal skin: melanin, 'melanoid', carotene, reduced haemoglobin and oxyhaemoglobin. 'Melanoid' was believed to be a degradation product of melanin, but because other studies (e.g. Buckley & Grum, 1961) have failed to demonstrate this, it will not be discussed.

Carotene produces an absorption band in the blue region of the spectrum (at 482 nm). Edwards & Duntley (1939b) maintained that carotene was important in skin pigmentation, the carotene being derived entirely from the diet. However, although carotenoids are present in the skin (the epidermis having more than the dermis) (Lee *et al.*, 1975) their concentrations are so minute that they can hardly be a significant determinant of skin colour – except in a condition called *carotenaemia*[2] which results in a yellow pigmentation of the skin (similar to jaundice, but without the discoloration of the sclerae). A study in full-term Caucasoid infants with high serum bilirubin levels (jaundice) found lower skin reflectances at 465 nm, a wavelength at which the bilirubin compound shows characteristic absorption (Schreiner *et al.*, 1979).

Reduced haemoglobin and oxyhaemoglobin show maximal absorption bands in the spectral region between 542 nm and 576 nm (green and yellow). In general, because the vascularity of the skin influences skin colour, those areas with a rich arterial supply (e.g. the nipples, the red part of the head and neck) have pronounced absorption bands of oxyhaemoglobin on reflectance spectrophotometry. Even the exertion of excessive pressure on the skin by the applicator head of the instrument can alter the haemoglobin content to produce reflectance variations in the green region (Lees, Byard & Relethford, 1978).

Melanin absorption characteristics

The graph in Fig. 7.2 show the spectral reflectance curves (obtained from an EEL machine) in groups of male and female Negroids (Fali Tinguelin), San (Bushmen) and Caucasoids (Dutch). The subjects were tested at an unexposed area (inner upper arm) and an exposed area (forehead). The slope of the curves is essentially determined by melanin, and it is evident that there is a steady increase in reflectances from the violet end of the spectrum (where there is marked absorption by melanin) to the red end (where there is much less absorption). The high skin content of melanin in the Negroid groups results in far lower reflectances throughout the visible spectrum than in the Caucasoid groups, while the moderately pigmented San have reflectance curves in an intermediate position. The influence of gender on skin colour will be discussed below but Fig. 7.2 illustrates that the Negroid and San females tend to have lower levels of pigmentation than their male counterparts.

Within each group of subjects the forehead reflectances are lower than those at the inner upper arm. The increased melanin concentrations at the forehead are due to chronic solar exposure. The arm reflectance:forehead reflectance ratio has been used as the *tanning index*, and this index is the most sensitive to variations among world populations where it is based on readings at wavelength 425 nm (Ducros, Ducros & Robbe, 1975). It is clear from Fig. 7.2 that the tanning index is much higher (i.e. there is proportionately more melanin at the forehead than at the arm) in the Caucasoids than in the San or the Fali Tinguelin – in fact, in the Fali the index approximates to unity (i.e. there is an absence of tanning).[3] Thus, subjects with a low level of natural pigmentation tan to a relatively larger degree than those of darker pigmentation. Furthermore, as the forehead reflectance curves show, females exhibit less tanning than males, probably because males are generally more involved in outdoor activities and have greater exposure to the sun.

In the Caucasoid reflectance curves there is a conspicuous dip between wavelengths 515 nm and 575 nm due to haemoglobin (particularly oxy-haemoglobin) absorption. This dip is absent from the San and Negroid curves because the presence of more abundant melanin in these populations has masked the underlying haemoglobin pigment. The spectral reflectance curves differ between Caucasoids and Negroids only in the region 300–1200 nm. Below and above these limits, variations in skin melanin concentration among differently pigmented people are not detected spectrophotometrically.

Although skin melanin concentrations are inversely related to skin reflectances the reflectances at different wavelengths are not linearly

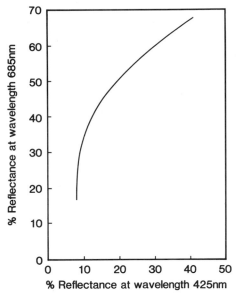

Fig. 7.3 Curvilinear relationship between reflectances at 425 nm and at 685 nm. Readings at 425 nm discriminate poorly among dark-skinned populations.

proportional to one another. There is, for example, a curvilinear relationship between reflectances at 425 nm and those at 685 nm (Fig. 7.3) (Weiner *et al.*, 1964; Rigters-Aris, 1973a). The plot indicates that among very dark-skinned subjects (i.e. those with reflectances at 425 nm of about 8 per cent) there is negligible variation at 425 nm, but considerable variation in the reflectances at 685 nm (which range from about 20 to 30 per cent). In contrast, among light-skinned subjects (i.e. those with reflectances at 425 nm of about 35 per cent) there is an almost linear relationship between the 425 nm and the 685 nm reflectances. Thus, the greater the degree of skin pigmentation the less discriminating are reflectances at 425 nm and the more discriminating are those at 685 nm. This wavelength variability is also demonstrated by the standard deviation of the reflectances from 425 nm up to 685 nm. In Negroids (see Table 7.1 note m) there is a steep increase in the standard deviation (i.e. in the amount of individual variation) across the spectrum, whereas in European Caucasoids (Table 7.4, note d) there is a decrease. Those of intermediate pigmentation (e.g. subjects from the north of India) show little change in standard deviation with increasing wavelengths (Table

7.4, note d). Harrison & Owen (1964) postulated that the scales which best remove environmental interaction and are therefore most appropriate for genetic analysis are the logarithm of reflectance at 425 nm, the (untransformed) reflectance at 545 nm and the antilogarithm of the 685 nm reflectance. However, these suggested transformations have not really proved to be of value (Byard, 1981).

It has been shown *in vitro* that the melanin concentration is linearly proportional to the reciprocal of the reflectance value (Harrison & Owen, 1964). This relationship is least precise at the short wavelengths (and especially for high concentrations of melanin), but at long wavelengths linearity obtains over a considerable range of melanin concentrations. Thus, reflectances at the red end of the spectrum tend to be more meaningful as a measure of skin melanin than those in the blue or green regions. Moreover, haemoglobin absorption at the red end of the spectrum is so low that reflectances at these wavelengths (655 nm and 685 nm) are practically uninfluenced by changes in blood supply to the skin, such as occur after the induction of redness by UV radiation (Lee & Lasker, 1959) or of bloodlessness *in vivo* by the application (ionophoresis) of noradrenaline into the skin (Harmse, 1964). For these reasons, reflectances at filter 609 (685 nm) of the EEL machine are regarded as the best index of melanin pigmentation. Some investigators have confined themselves to this wavelength in their surveys on skin colour although, more often, those wanting an abridged method select three wavelengths, namely 425 nm (filter 601), 545 nm (filter 605) and 685 nm (filter 609).

Factors influencing skin reflectances
Age

Post *et al.* (1976) measured the skin reflectances in healthy Caucasoid and Negroid infants who were between 25 and 44 weeks of gestational age: the babies were tested within 48 hours of birth. Negroid and Caucasoid newborns exhibited similar reflectances until 32 weeks of gestational age, after which the Caucasoids became lighter skinned and the Negroids darker. It is of interest that the foreheads of the Caucasoid and Negroid babies were respectively 10 and 12 per cent darker than the inner aspect of the upper arms. Similarly, in a reflectance study of New Guinea newborn infants (Walsh, 1964) the forehead readings were lower than those from the forearm and chest wall. The reflectance difference between forehead and upper arm is usually taken as a measure of the so-called 'tanning capacity' of the skin, but the presence of such a difference at birth does detract from the specificity of the measure. As already discussed (see p. 6), the melanocyte density of the forehead is among the highest of any bodily region. It is comparable with that in the genitalia, although the scrotum of the New Guinea infant was in fact far more pigmented than the forehead. The most striking aspect of Walsh's study is the rapidity with which the infant darkened: by the age of 6 months the skin had attained the melanin content of the young adult.

Studies of Quecha Indians (Conway & Baker, 1972), Peruvian Mestizos (Frisancho, Wainwright & Way, 1981) and Sikhs (Kahlon, 1973) have found that in the unexposed areas of both sexes, skin reflectances gradually tended to decrease from childhood until early adolescence and to increase thereafter and during adulthood. The phenomenon of skin darkening at around puberty has been ascribed to the increased secretion of pituitary hormones at this period of life (Conway & Baker, 1972). The pineal hormone, melatonin (see p. 34), has inhibitory actions on sexual development, and blood levels of melatonin decrease from early childhood to young adulthood, reflecting the development of sexual maturation (Waldhauser *et al.*, 1984). This decline in melatonin may possibly act as the trigger for the increased skin pigmentation at puberty (Silman *et al.*, 1979). Surveys among the Ainu of Japan (Harvey & Lord, 1978), the Fali of Cameroon (Rigters-Aris, 1973a) and the Sara Madinjay of Chad (Hiernaux, 1972) have found no consistent association between skin reflectances and age.

With aging, therefore, there is generally either no significant change in the reflectances of the unexposed skin or a trend towards increased reflectances. The latter would be in keeping with the 8–20 per cent fall-off of melanocytes per decade of life (see p. 22). However, the situation is different with the exposed skin: here increasing age causes decreased reflectances (Walsh, 1964), even though there is a comparable decline in melanocyte numbers at exposed sites. The explanation appears to be that the cumulative solar exposure stimulates the melanocytes and, while their number has diminished, their capacity for melanin formation has increased to the extent that there is a net increase in skin pigmentation (Gilchrest *et al.*, 1979).

Longitudinal studies of American Caucasoid twins (identical and fraternal) from 3 months to 6 years of age showed a darkening with age of both the hair colour (Matheny & Dolan, 1975a) and the eye colour (Matheny & Dolan, 1975b), the darkening process being more marked and occurring earlier in females. Furthermore, a very high rate of concordance for hair and eye colour was found in the identical (but not in the fraternal) twins at every age, indicating a strong genetic influence in the timing of these colour changes.

Table 7.1. *Reflectance readings taken with the EEL Reflectance Spectrophotometer at the inner (medial) aspect of the upper arm: Africa*

(Populations ranked in order of increasing reflectance at 685 nm)

Locality	Population[a]	Sex	Age (yrs)	Number	Percentage reflectance (mean) Filters (Dominant wavelength in nm)									Reference
					601 (425)	602 (465)	603 (485)	604 (515)	605 (545)	606 (575)	607 (595)	608 (655)	609 (685)	
Mozambique Sofala	Chopi	M	adults	105	7.3								17.8	Weninger (1969)
		F	adults	91	7.7								21.1	
Cameroon (Northern)	Fali Tinguelin	M	all ages	120	7.3	7.4	7.4	7.6	7.9	8.5	11.3	16.1	20.0[m]	Rigters-Aris (1973a)
		F	from 6[b]	43	7.7	7.6	7.6	7.9	8.2	8.8	12.0	16.9	20.9	
	Fali Kangou	M	all ages	18	7.4	7.7	7.7	8.0	8.4	8.9	11.9	16.8	21.1	
		F	from 11[b]	22	8.1	8.0	8.0	8.5	8.7	9.3	12.6	18.6	25.5	
Namibia (bordering on Angola)	Okavango Bantu,[c] Kuangali[d,e]	M	adults	50									21.6	Weiner et al. (1964)
		F	adults	38									22.1	
		M	children	102									24.2	
		F	children	144									25.6	
	Okavango Bantu,[c] M'Bukushu at Bagani[e]	F	adults	23									22.2	
		M	adults	15									22.6	
		M	children	21									22.4	
		F	children	19									25.5	
Chad	Sara Madjingay from Ndila	M	all ages[b]	415									23.2	Hiernaux (1972)
		F	all ages[b]	359									26.0	
Nigeria (Southern) Ibadan	Yoruba[f]	M	18–60	100	8.0	8.3	8.1	9.7	10.1	10.9	14.3	20.6	23.6[m]	Barnicot (1958)
		F	18–60	94	8.5	8.9	8.9	10.5	11.1	12.2	16.2	23.1	26.1	
Tanzania	Nyatura	M	not stated	not stated	7.8				10.2				25.3	Weiner (1969)
		F	not stated	not stated	8.6				10.6				26.3	
Malawi	Mainly Cewa[g]	M	not stated	38	9.0	10.2	9.5	9.8	10.0	11.9	16.4	23.3	27.0	Tobias (1974)

Country/region	Group	Sex	N	Age										Reference
Burkina Faso (formerly Upper Volta)	Kurumba[h] from Roanga	F	149	all ages[j]	10.8	12.0	11.0	11.0	11.6	13.0	17.1	27.1	27.9	Huizinga (1968)
		M	412	all ages[j]	11.4	12.5	11.6	11.4	12.0	13.5	17.4	28.0	29.3	
Tanzania	Sandawe	M	not stated	not stated	8.8				11.1				28.1	Weiner (1969)
		F	not stated	not stated	9.1				11.7				29.7	
Nigeria (Southern) Ibadan	Ibo[f]	M	52	students	3.7	9.2	9.3	10.7	11.2	12.6	17.0	24.6	28.2	Barnicot (1958)
Namibia (bordering on Angola)	Black Bushmen[c] at Bagani[i]	M	13	adults									28.2	Weiner et al. (1964)
		F	25	adults									29.4	
		M	14	children									28.0	
		F	14	children									32.4	
Zaire	Congolese (except for 3 females from Cameroon)	M	73	20–40	7.4	8.6	9.0	9.7	10.4	13.3	18.2	25.6	29.6	Van Rijn-Tournel (1966)
		F	36	13–39	9.0	10.7	11.5	12.5	12.8	15.4	21.7	30.1	36.1	
Ethiopia	Residents of Adi-Arkai (1500 m altitude)	M	55–61[g]	adults	8.3				11.0				30.2	Harrison et al. (1969)
		F	25–34[g]	adults	9.8				13.1				33.2	
	Residents of Debarech (3000 m altitude)	M	43–42[g]	adults	9.5				12.7				31.4	
		F	13	adults	10.3				14.3				35.7	
South Africa	Bantu[c-e] (96 per cent Xhosa)	M	104	14–54	10.6	12.3	12.4	12.8	13.7	16.4	22.5	30.1	32.1	Wassermann & Heyl (1968)
		F	100	17–50	12.6	14.5	14.8	15.5	16.8	19.7	26.5	35.0	38.9[m]	
South Africa (Johannesburg) & Swaziland	Swazis[e]	M	47	not stated	9.7	11.0	10.3	13.3	13.9	14.5	18.4	27.6	32.4	Roberts, Kromberg & Jenkins (1986)
		F	76	not stated	10.5	12.2	12.3	14.9	16.5	18.1	22.7	32.8	38.8[m]	
Nigeria Lagos	Africans	M	138	8–40	12.8	12.6	13.1	14.1	14.9	16.9	22.5	28.3	32.5	Ojikutu (1965)
Mali	Dogon[h] from Sanga and Boni	M	121	all ages[j]	[k]	11.5	11.0	11.4	11.9	13.7	18.2	32.4	33.7	Huizinga (1968)
		F	104	all ages[j]	[k]	11.8	11.2	11.5	11.9	14.0	18.5	33.5	34.5	
Botswana Kalahari Desert	Yellow Bushmen[c] at Lone Tree	M	25	adults	14.4								40.5	Weiner et al. (1964)
		F	28	adults	15.2								43.1	
South Africa Johannesburg	South African Negroes (73 per cent Tswana and Xhosa)[e]	M	40	19–66	15.3	17.9	17.0	18.1	18.2	20.7	27.8	39.0	41.7	Robins (1972)
		F	43	23–50	16.5	19.3	18.7	20.0	20.6	23.4	31.2	41.5	44.3	

Table 7.1. Continued

Locality	Population[a]	Sex	Age (yrs)	Number	Filters (Dominant wavelength in nm) 601 (425)	602 (465)	603 (485)	604 (515)	605 (545)	606 (575)	607 (595)	608 (655)	609 (685)	Reference
Namibia Warmbath	Hottentot[c]	M	adults	25									41.9	Weiner et al. (1964)
		F	adults	50									45.6	
Botswana Kalahari Desert	Central Bushmen[c]	M	adults	42	12.7	15.5	16.2	17.2	18.1	21.1	28.4	39.4	42.5	Tobias (1961)
		F	adults	46	13.7	16.6	17.4	18.2	19.0	22.5	30.2	40.8	43.6	
	Yellow Bushmen[c] at Takashwani	M	adults	20	16.2								43.0	Weiner et al. (1964)
		F	adults	19	17.2								43.6	
	Yellow Bushmen[c] at Ghanzi	M	adults	43	15.6								43.0	
		F	adults	12	16.4								44.6	
South Africa Namaqualand	Hottentot[c]	M	adults	25									45.5	Weiner et al. (1964)
		F	adults	34									48.1	
Namibia Rehoboth	Basters	M	not stated	10	17.2								47.9	Weiner et al. (1964)
		F	not stated	21	20.6								51.9	
South Africa Cape Province	Cape Coloureds[e]	M	16–65	103	15.8	19.4	20.5	21.5	22.7	26.8	36.1	45.2	49.2	Wassermann & Heyl (1968)
		F	16–46	107	18.5	22.4	24.3	25.4	27.0	31.0	40.3	48.6	52.1	
	Cape Coloureds[e]	M	not stated	187									50.1	Weiner et al. (1964)
		F	not stated	112									51.3	
Algeria Aures	Chaouias from Bouzina[f]	M	16–57	114									57.4	Chamla & Demoulin (1978)
		F	16–57	115									58.7	
South Africa	Whites	M	16–62	108	29.5	35.9	37.4	37.9	37.4	42.1	53.7	60.9	63.5	Wassermann & Heyl (1968)
		F	17–41	109	32.3	38.0	40.2	41.0	40.2	45.0	56.4	62.6	64.4	
	Caucasians	M	17–65	40	35.4	40.5	42.1	42.6	39.9	43.8	55.9	64.4	65.8	Robins (1972)
		F	16–61	40	36.5	41.0	42.7	43.2	41.7	45.1	57.2	64.6	65.9	

[a] The description of the population sample is directly taken from that used by the original author(s).

[b] The original paper also gives a breakdown of skin reflectances according to age categories.

[c] These terms have been generally replaced in the scientific literature by the following: South African or Namibian Negro/Negroid for *Bantu*, San for *Bushmen* and Khoikhoi for *Hottentot*.

[d] The mean reflectances cited here represent pooled data from samples measured at Mazua, Kakuru, Kurungkuru and Tondoro.

[e] The marked difference in skin reflectances between the Namibian and South African Negroid groups can be attributed to the varying degrees of Khoisan (Hottentot-Bushman) admixture in the latter. According to the estimates of Jenkins, Zoutendyk & Steinberg (1970), the Xhosa and Tswana have about 60 per cent Khoisan admixture and the Swazis 25 per cent whereas the Namibian/ Angolan groups have no Khoisan admixture. (A greater degree of Khoisan admixture is likely to be associated with a lighter skin colour, i.e., higher reflectances.) The 'Cape Coloureds' are derived from the Khoikhoi, San, Malays, Negroids and European settlers.

[f] The reflectance readings on these populations were taken not on the inner upper arm but on the flexor surface of the forearm at the junction of the middle and proximal thirds.

[g] The mean reflectance values cited here are not necessarily accurate as they are projected from a skin reflectance graph which is the only form in which the data are presented in the original paper.

[h] In these studies a 'white' enamelled plate was used as the standard instead of the usual 'white' magnesium carbonate block and this has resulted in slightly lowered reflectances.

[i] The original paper cited reflectances only in terms of age categories but the data have been pooled to give mean reflectances for the total sample.

[j] A range is quoted because, whereas all the subjects were measured on filter 609, varying numbers were measured on the other filters.

[k] Readings were not taken at this wavelength because of a defective 601 filter.

[l] In this study, one group of subjects (male and female) was tested on the inner upper arm in September–October and the other in May–June: the former had higher reflectances (highly significant in males), thus demonstrating seasonal differences even at an apparently unexposed skin area.

[m] The standard deviations of the reflectances at filters 601 to 609 respectively are cited below for the following population samples: Fali Tinguelin (males): 1.0, 1.0, 1.1, 1.1, 1.1, 1.2, 1.3, 2.2, 3.3 and 3.6; Yoruba (males): 0.9, 1.0, 1.2, 1.4, 1.2, 1.5, 3.1, 3.5 and 3.8; Xhosa (females): 2.0, 2.4, 2.6, 2.6, 2.9, 3.4, 4.4, 4.9 and 4.5; Swazis (females): 1.7, 2.3, 2.3, 2.8, 3.0, 3.2, 3.9, 5.0 and 5.1.

Sex

Most of the skin reflectance studies carried out throughout the world have investigated females and males separately, and the data are presented in Tables 7.1–7.4. The overwhelming tendency is that females have higher reflectances (i.e. are lighter in skin colour) than males. This observation embraces populations as geographically diverse as the English in Northumberland (Hulse, 1973), the San of the Kalahari (Tobias, 1961), the Yoruba of Nigeria (Barnicot, 1958), the Papuans of New Guinea (Harvey, 1985), the Japanese (Hulse, 1967), and the *Garifuna* and Creoles of Central America (Byard & Lees, 1982). Several studies (Huizinga, 1968; Kalla & Tiwari, 1970; Kalla, 1973) have noted that skin pigmentation in the unexposed body areas shows a sex reversal at adolescence: in the pre-adolescent period girls are darker than boys, but at ages 13–16 years the boys become darker. This pigmentary change possibly relates to the differential hormonal status of the sexes at adolescence.

There are, however, exceptions to the general trend of a lighter skin colour in females. Some European surveys (e.g. of the Dutch (Rigters-Aris, 1973b) and of the Belgians (Leguebe, 1961)) have revealed higher reflectances in males at the inner upper arm (see Table 7.4), and the same was found among the Kurumba of Burkina Faso (formerly Upper Volta) (Huizinga, 1968) (see Table 7.1). At the exposed areas (such as the forehead) males are invariably more pigmented than females, even in those cases where they tend to be less pigmented than females in the unexposed sites. The pronounced male–female difference in the exposed skin (Fig. 7.2) appears to be due to greater sun exposure (and thus tanning) in males. Whereas the inner aspect of the upper arm has been selected as representing an area of untanned skin – and therefore of natural (genetic) pigmentation – this may not be a justifiable assumption. Lasker (1954) has demonstrated that reflectances at the inner upper arm drop in parallel (but to a lesser extent) with those at the forehead during the summer months and then rise again with them in winter. Other seasonal studies (Chamla & Demoulin, 1978; Little & Wolff, 1981) have also shown that the inner upper arm may not be exempt from tanning.[4] Among the Ainu of Northern Japan the males are markedly darker than the females at the inner upper arm (Harvey & Lord, 1978) (see Table 7.3). This sexual dimorphism is explained by clothing patterns – the men frequently wear short-sleeved shorts or sleeveless singlets whereas women don long-sleeved garments which cover the upper arm. Variations in clothing styles and in solar exposure (dictated by social and religious custom) probably contribute substantially to the vast gender differences in inner upper arm reflectances among Northern Iranians (see Table 7.3) (Mehrai & Sunderland, 1990). Conversely, the darker skin colour of Kurumba females at the upper arm (Huizinga, 1968) may be due to their practice of carrying water pots on their heads with one arm raised. Thus, if there are sex difference in skin reflectances at the 'unexposed' arm site, then these may be associated less with genetic and physiological effects than with cultural habits and occupational activities.

Social determinants

Skin reflectances may also be influenced by social factors. In Japan social selection has favoured the concentration of alleles for lighter skin colour in the upper classes of society (Hulse, 1967). In India, high-caste–low-caste differences in skin reflectances are evident (Table 7.3, note b), with the higher castes having higher reflectances (Das & Mukherjee, 1963).

A related area in which skin colour may have the attributes of a status symbol is in mate selection. Reflectance studies have demonstrated the phenomenon of *assortative mating* for skin colour, i.e. that marital partners in a particular community so resemble each other in pigmentation that this characteristic *per se* directly modifies mating behaviour in that society. Positive assortative mating for colour (i.e. high husband–wife correlations for skin reflectances) has been noted in Punjabi Indians (Banerjee, 1985), Sikhs (Roberts & Kahlon, 1972), Brazilian Negroids (Harrison *et al.*, 1967), Solomon Islanders (Baldwin & Damon, 1973) and Peruvian Mestizos (Frisancho *et al.*, 1981).

Hair and eye colour

The extent to which hair and eye coloration are correlated with skin colour has not been extensively studied. Little & Wolff (1981) measured skin and hair reflectances in young women with red hair. They found that skin reflectances at the inner upper arm fell at the higher end of the range of variation for north-western European women, thus confirming the light complexion of redheads. The hair and inner arm reflectance values were highly correlated at the short wavelengths: this suggested that phaeomelanin was the source of the relationship. Furthermore, when the redheads were divided into those with light eyes (blue, green and blue-green) and those with dark eyes (brown and hazel) the former tended to have higher hair and skin reflectances than the dark-eyed subsample. A previous study (Robins, 1973) showed that blue-eyed females (but not males) had significantly higher reflectances at the inner upper arm than brown-eyed females but at a sun-exposed hand area the effects of tanning had markedly reduced these differences.

Table 7.2. *Reflectance readings (EEL Reflectance Spectrophotometer) at the inner upper arm: Americas*

(Populations ranked in order of increasing reflectance at 685 nm)

Locality	Population[a]	Sex	Age (yrs)	Number	Percentage reflectance (mean) Filters (Dominant wavelength in nm)									Reference
					601 (425)	602 (465)	603 (485)	604 (515)	605 (545)	606 (575)	607 (595)	608 (655)	609 (685)	
Central America														
Belize	*Garifuna*[b]	M	all ages but mainly 1–15	99	7.2	8.0	8.7	9.2	10.3	12.3	17.1	23.4	27.6	Byard & Lees (1982)
		F		209	7.2	8.1	8.9	9.5	10.8	13.0	18.2	24.3	28.8[f]	
	Creoles[b]	M	all ages but mainly 1–15	70	7.9	9.0	10.1	10.5	12.1	14.5	20.3	27.3	31.4[f]	
		F		105	8.2	9.3	10.2	10.9	12.6	15.1	20.9	27.9	32.0	
North America (USA)														
Arizona (McNary)	Black Americans	M	18–81	49	11.3	13.3	13.8	14.4	15.4	18.0	24.7	31.7	34.6[f]	Wienker (1979)
South Carolina	American-born blacks[c,d]	M	18–27	197	13.7		14.7		16.7		25.5		35.9	Spurgeon, Meredith & Onuoha (1984)
Peru														
Marañon Valley	Aguarana Indians	M	5–60	40					24.5				42.8	Weiner, Sebag-Montefiore & Peterson (1963)
		F	5–60	52					24.7				43.3	

Location	Population	Sex	Age	N						Reference
Brazil										
Porto Alegre	Southern Brazilian Negroes[g]	M	7–16	68	14.9		19.9		42.9	Harrison et al. (1967)
		F	7–16	117	17.1		22.9		46.0	
Paraná	Guarani Indians	F	7–50	10	17.2		23.4		46.5	Harrison & Salzano (1966)
		M	3–65	23	16.8		23.6		47.9	
	Caingang Indians	M	2–70	55–60[e]	16.5		22.9		48.1	
		F	3–65	34–40[e]	18.4		26.1		50.7	
Greenland East										
Ammassalimiut	Eskimo	M	mainly adults up to 55, some adolescents	102	22.4		29.9		55.7	Ducros, Ducros & Robbe (1975)
Brazil										
Porto Alegre	Southern Brazilian 'whites'	M	7–16	216	25.5		33.6		57.9	Harrison et al. (1967)
		F	7–16	258	28.2		35.5		59.3	
North America (USA)										
South Carolina	American-born whites[c,d]	M	18–27	97	35.6	41.8	41.8	57.1	63.9	Spurgeon, Meredith & Onuoha (1984)

[a] The description of the population sample is directly taken from that used by the original author(s).

[b] It is estimated that the *Garifuna* gene pool consists of 75 per cent African derivation, 22 per cent Amerindian and 3 per cent European, and the Creole 72 per cent European, 8 per cent and 20 per cent respectively (Byard & Lees, 1982).

[c] The reflectance spectrophotometer used in this study was not the EEL but a similar machine with comparable filters.

[d] The original study also measured reflectances in subgroups of subjects classified according to grandparental ancestry.

[e] A range is quoted because, whereas all the subjects were measured on filter 609, varying numbers were measured on the other filters.

[f] The standard deviations of the reflectances at filters 601 to 609 respectively are cited below for the following population samples: *Garifuna* (females): 1.5, 1.7, 2.0, 2.1, 2.5, 2.9, 3.9, 4.9 and 4.8; Creoles (males): 2.1, 2.5, 3.2, 3.4, 3.9, 4.6, 5.8, 7.1 and 7.3; Black Americans (males): 3.6, 4.4, 4.9, 5.3, 4.8, 5.5, 6.5, 7.8 and 8.1.

[g] Between 40 and 50 per cent of the skin colour genes in the Brazilian Negro are of European origin and there is evidence of positive assortative mating for skin colour in the Brazilian Negro (Harrison et al., 1967).

Table 7.3. *Reflectance readings (EEL Reflectance Spectrophotometer) at the inner upper arm: Asia and Australia*

(Populations ranked in order of increasing reflectance at 685 nm)

Locality	Population[a]	Sex	Age (yrs)	Number	Filters (Dominant wavelength in nm)									Reference
					601 (425)	602 (465)	603 (485)	604 (515)	605 (545)	606 (575)	607 (595)	608 (655)	609 (685)	
Papua-New Guinea	Lufa Villagers	M	adults over 20	189	9.1				13.3				30.8	Harvey (1985)
		F	20	229	9.3				13.8				31.6	
	Karker Islanders	M	adults over 20	217	9.6				13.8				30.8	
		F	20	256	10.1				15.0				33.2	
India Koraput town, Orissa	Bareng Paroja[b]	M	adults	100	9.8	10.3	10.7	11.5	12.4	14.6	20.0	28.8	31.8[f]	Das & Mukherjee (1963)
	Bado Gadaba[b]	M	adults	100	10.3	10.7	11.2	11.8	12.7	15.2	20.7	29.7	32.3	
Nagpur & Kamptee	Mahar[b]	M	adults	100	13.3	14.8	15.7	16.9	18.2	21.6	28.4	37.6	41.3	Büchi (1957)
Bengal	Low caste[b]	M	adults	40	13.9	17.0	18.2	19.9	20.4	24.1	29.9	41.5	44.8	Kalla (1972)
Delhi	Saxena Kayastha (husband–wife–children)	F	adults	8	13.9	17.1	18.9	19.6	21.5	24.8	32.8	41.5	45.8	
		M	adults	8	16.8	20.3	22.8	23.5	24.8	29.0	37.6	46.5	50.4	
		M+F	children	22	14.5	18.0	19.9	20.9	22.3	26.0	34.4	43.7	48.0	
Jordan (Eastern) Azraq	Arab villagers	F	mean: 8.3	3	16.7				24.2				48.0	Sunderland (1967)
		M	mean: 15.2	12	20.3				27.6				51.5	
India North India	Baniya[b]	M	20–25	104	14.3				22.0				48.6	Jaswal (1979)
Bengal	Kayastha[b]	M	adults	58	16.3	19.7	21.1	22.8	24.3	27.4	35.6	45.3	48.6	Büchi (1957)
Delhi	'Aggarwal'	M	10–16[c]	414									49.5[d]	Kalla (1973)
		F	10–16[c]	371									50.0[d]	
Calcutta	Rarhi Brahman[b]	M	adults	100	18.3	20.3	21.9	23.2	24.5	29.1	36.9	47.7	49.7[f]	Das & Mukherjee (1963)
Bengal	Brahman[b]	M	adults	76	17.2	20.4	21.7	24.0	24.3	28.5	36.3	47.5	49.9	Büchi (1957)
Iran (Northern)	Nowshahr City	M	8–24[h]	68	18.2	21.6	24.4	26.6	27.2	31.2	39.7	46.5	50.0	Mehrai & Sunderland (1990)
		F	8–24[h]	33	28.0	32.9	35.6	37.3	37.6	41.9	50.9	56.6	59.7	

Location	Population	Sex	Age	N										Reference
India														
North India	Jat Sikhs[b]	M	20–25	101	15.4				23.6				50.6	Jaswal (1979)
Bengal	Vaidya[b]	M	adults	19	18.4				25.3				50.7	Büchi (1957)
North India	Haryana Jats[b]	M	20–25	97	15.7				24.0				51.5	Jaswal (1979)
Japan (South-West)	Japanese[e]	M	15–19	116	f	23.0	24.3	25.7	26.9	31.7	41.6	49.0	51.6	Hulse (1967)
		F	15–19	68	f	29.7	31.4	31.0	33.7	37.9	47.0	53.4	55.5	
Jordan (Eastern) Azraq	Chechen	F	mean: 9.5	7	21.2				29.1				51.9	Sunderland (1967)
		M	mean: 20.4	23	27.4				33.6				55.0	
India (North)	Rajputs[b]	M	20–25	70	16.2				24.7				52.0	Jaswal (1979)
	Khatris[b]	M	20–25	176	16.5				25.1				52.7	
Israel Baraket	Yemenite Jews from Habban	M	all ages but 60% aged 16 or under	449 (M+F)	19.7	23[i]	24	25	27	31	40	49	52.3	Hulse (1969)
		F			22.9	i							56.7	
Jordan (Eastern) Azraq	Non-village Arabs	M	mean: 27	8	22.4				30.4				52.7	Sunderland (1967)
India (North)	Brahmans[b]	M	20–25	102	27.2				26.0				52.8	Jaswal (1979)
Jordan (Eastern) Azraq	Druze	M	mean: 17	42	22.5				30.0				52.8	Sunderland (1967)
Japan (Central)	Japanese[e]	M	15–19	54	f	24.4	26.3	27.0	27.7	33.0	42.8	50.5	53.3	Hulse (1967)
		F	15–19	51	f	28.6	30.4	32.1	32.9	37.1	46.4	53.2	55.7	
India Delhi	Punjabi Indians (father–mother–son–daughter)	F[g]	adults	309	17.9	22.0	24.4	25.9	26.2	31.2	40.5	49.7	53.6	Banerjee (1984)
		M[g]	adults	309	18.8	23.3	25.5	27.0	27.0	31.8	41.4	50.8	54.5	
		M	children & adolescents	252	17.8	22.4	24.5	26.0	26.2	30.9	40.6	50.0	54.3	
				190	18.6	23.0	25.5	27.0	27.2	32.3	41.8	51.0	54.9	
Japan (North)	Japanese[e]	M	15–19	54	f	26.1	27.8	28.8	29.5	34.0	44.1	51.4	54.1	Hulse (1967)
		F	15–19	50	f	29.7	31.2	33.1	33.6	37.9	46.8	53.5	55.7	
India Mussoorie	Tibetans	M	12–18[c]	135									54.5	Kalla & Tiwari (1970)
		F	10–16[c]	135									54.9	
North India	Aroras	M	not stated	171	20.7								55.8	Kalla (1969)
		F	not stated	209	21.1								56.1	
	Khatris	M	not stated	249	21.3								56.1	
		F		238	21.7								56.4	

Table 7.3. *Continued*

Locality	Population[a]	Sex	Age (yrs)	Number	Filters (Dominant wavelength in nm)									Reference
					601 (425)	602 (465)	603 (485)	604 (515)	605 (545)	606 (575)	607 (595)	608 (655)	609 (685)	
Japan Hidaka, Hokkaido	Ainu	M	10–60[h]	50	f	29.2	30.8	31.9	31.7	35.6	46.0	54.3	57.4	Harvey & Lord (1978)
		F	10–60[h]	101	f	33.3	35.1	36.8	37.6	41.0	51.1	58.2	60.8[i]	
India Calcutta	Europeans	M	adults	10	28.5	37.5	40.0	39.3	38.8	41.3	53.5	61.5	64.3	Büchi (1957)

[a]The description of the poulation sample is directly taken from that used by the original author(s).
[b]There is a positive relationship among these Indian population samples between skin reflectances and caste/tribe such that the higher castes tend to be of lighter skin colour.
[c]The original paper cited reflectances only in terms of age categories but the data have been pooled to give mean reflectances for the total sample.
[d]The original paper expressed reflectances at this wavelength as 'antilogarithms'.
[e]There is a positive relationship among these Japanese subjects between skin reflectances and social class.
[f]Readings were not taken at filter 601 because they were unreliable with the electric current available in Japan.
[g]In a subsequent paper, Banerjee (1985) showed that positive assortative mating for skin colour was evident among these spouse pairs.
[h]The original paper also gives a breakdown of skin reflectances according to age categories.
[i]The author gives only approximate readings in males for filters 602–608 and states that female reflectances at those wavelengths are always between 4 and 5 points higher than the male.
[j]The standard deviations of the reflectances at filters 601 to 609 respectively are cited below for the following population samples: Bareng Paroja (males): 1.2, 1.5, 1.7, 1.6, 1.4, 2.3, 3.0, 3.6 and 3.9; Rahri Brahman (males): 4.2, 4.7, 5.3, 5.5, 5.2, 5.0, 6.0, 6.9 and 7.0; Punjabi Indians (females): 3.2, 3.8, 4.2, 3.4, 3.2, 4.1, 5.1, 4.9 and 4.1; Ainu (females): —, 4.7, 5.1, 4.8, 4.5, 4.8, 5.2, 4.0 and 3.7.

Table 7.4. *Reflectance readings (EEL Reflectance Spectrophotometer) at the inner upper arm: Europe*

(Populations ranked in order of decreasing reflectance at 685 nm)

Locality	Population[a]	Sex	Age (yrs)	Number	Percentage reflectance (mean) Filters (Dominant wavelength in nm)									Reference
					601 (425)	602 (465)	603 (485)	604 (515)	605 (545)	606 (575)	607 (595)	608 (655)	609 (685)	
England	North Northumberland	F	15–16	104	36.5	42.1	45.8	46.4	44.8	50.0	59.6	66.7	68.9	Hulse (1973)
		M	15–16	93	34.2	40.0	43.8	45.3	43.8	48.9	58.7	66.2	68.6	
Netherlands	Dutch (mainly resident in Utrecht)	M	all ages[b]	99	39.7	46.8	48.3	47.8	44.7	49.0	60.9	67.5	68.9[d]	Rigters-Aris (1973b)
		F	all ages[b]	100	37.7	44.6	45.8	45.8	43.2	47.4	58.7	65.3	66.7	
England	South-East Northumberland	F	15–16	51	35.6	41.1	44.6	45.5	43.9	48.9	59.1	65.9	68.3	Hulse (1973)
		M	15–16	55	33.0	38.9	42.7	44.0	42.1	47.4	56.8	64.4	66.8	
Belgium	Brussels	M	university students	143	37.7	44.7	46.7	46.7	44.8	48.9	59.5	65.7	67.3	Leguebe (1961)
		F		177	36.5	43.5	45.3	45.7	44.6	48.8	58.5	64.3	65.9	
England	Cumberland	F	secondary schoolchildren	153	36.9				42.4				67.0	Smith & Mitchell (1973)
		M		99	35.8				41.8				66.5	
British Isles	Isle of Man	F	secondary schoolchildren	73	36.8				41.8				67.0	Smith & Mitchell (1973)
		M		90	36.6				41.9				65.9	
Germany Mainz	German & American whites	M	19–20	74	40.3	45.5	47.4	47.1	45.2	50.1	60.6	66.7	66.9	Ojikutu (1965)
Spain Guipúzcoa	Basques	F	19–65	239	29.2	35.9	40.9	40.9	40.9	45.6	56.0	64.5	66.4	Rebato (1987)
		M	19–65	183	30.0	37.0	41.2	40.9	39.6	44.1	55.0	63.4	65.5	
León	Meseta	M	adolescents and adults to age 50	108–173[c]	29.0	36.4		39.0	36.8			62.8	66.1	Caro (1980)
Vizcaya	Basques	F	19–65	153	29.0	35.5	40.1	40.4	40.0	44.9	55.2	63.9	65.8	Rebato (1987)
		M	19–65	169	30.0	36.6	41.1	40.7	39.6	44.5	55.0	63.3	65.3	

Table 7.4. *Continued*

Locality	Population[a]	Sex	Age (yrs)	Number	Percentage reflectance (mean) Filters (Dominant wavelength in nm)									Reference
					601 (425)	602 (465)	603 (485)	604 (515)	605 (545)	606 (575)	607 (595)	608 (655)	609 (685)	
Ireland (Eire)	Ballinlough	M	all ages	105	35.4				40.9				65.3	Sunderland *et al.* (1973)
		F	all ages	127	36.2				41.9				65.1	
	County Longford	M+F	4–14	320	36.7	41.0	44.8	44.6	41.4	46.2	58.2	63.1	64.9[d]	Lees, Byard & Relethford (1979)
	Rossmore	F	all ages	90	35.7				41.8				64.8	Sunderland *et al.* (1973)
		M	all ages	111	34.6				40.7				64.7	
	Carnew	F	all ages	162	37.2				42.1				64.6	Sunderland *et al.* (1973)
		M	all ages	105	34.9				39.4				64.4	
Spain León	Cabrera	M	adolescents and adults to age 50	108–125[c]	26.3	33.4		36.5	36.1			61.0	64.6	Caro (1980)
Belgium Brussels	Belgians (except for 2 Greeks and 1 Iraqi)	M	18–35	69	39.1	45.5	46.6	46.1	43.8	47.5	57.4	62.9	64.5	Van Rijn-Tournel (1966)
		F	18–45	46	38.2	43.7	44.5	44.8	43.2	47.2	57.3	62.6	63.7	
Spain León	Bierzo	M	adolescents and	110–211[c]	27.4	34.4		37.1	35.5			60.6	64.3	Caro (1980)
	Montana	M	adults to age 50	117–146[c]	28.5	36.0		37.8	35.1			60.4	64.0	
Germany Mainz	Syrians & Iranians	M	23–27	6	38.9	41.2	42.7	43.3	43.0	46.7	57.7	62.3	64.0	Ojikutu (1965)
Wales	Merthyr Tydfil	F	secondary	98	33.2				38.7				63.5	Smith & Mitchell (1973)
		M	schoolchildren	84	32.8				38.7				62.8	
England Northumberland	Holy Island	M	adults[b]	49	34.1				39.8				63.4	Cartwright (1975)
		F	adults[b]	51	34.1				39.6				62.3	
Wales (South-West)	North Pembrokeshire	F	7–18[b]	148	35.6				40.9				63.1	Sunderland & Woolley (1982)
		M	7–18[b]	148	34.2				40.8				62.9	

Country, Place	Population	Sex	Age	N										Reference
England London	Europeans[e]	F	adults	50	34.3	41.4	43.1	43.9	40.5	43.9	54.3	61.5	63.1[d]	Barnicot (1958)
		M	adults	50	32.8	39.7	41.4	41.6	37.9	41.4	52.4	59.9	61.5	
Wales (South-West)	South Pembrokeshire	F	7–18[b]	224	34.1				40.5				62.7	Sunderland & Woolley (1982)
		M	7–18[b]	225	33.7				39.6				62.4	
Spain León	Maragateria	M	adolescents and adults to age 50	102–111[c]	25.1	31.8		34.8	33.4			59.5	62.7	Caro (1980)
England Liverpool	Europeans	M+F	adults	46–105[c]	36.1	41.6	43.0	43.7	41.0	45.2	54.8	61.7	62.3	Harrison & Owen (1964)
England	Sikhs (husband–wife–children)	F	adults[b]	30	23.4	27.5	28.3	30.5	30.9	36.5	45.3	52.0	55.6	Kahlon (1973)
		M	adults[b]	30	23.1	27.0	27.7	29.5	29.7	34.8	43.5	50.6	54.5	
		F	over 90% under 18	152	21.8	26.0	26.7	28.7	29.3	34.6	43.1	50.4	54.1	
		M		173	20.5	24.2	25.0	27.0	27.4	32.6	41.0	48.4	52.4	
England London	North Indians (higher castes)	M	adults	153	21.4	25.2	26.5	28.0	27.9	31.5	40.5	49.4	52.5[d]	Tiwari (1963)
Germany Mainz	American Negro	M	19–26	12	19.3	18.6	18.5	21.2	22.8	25.9	33.9	39.3	43.9	Ojikutu (1965)
England London	West Africans	M	adults	37–106[c]	12.3	13.2	13.4	14.6	14.4	16.6	21.7	29.9	34.7	Harrison & Owen (1964)
Germany Mainz	Africans from Ghana & Liberia	M	22–40	13	13.6	12.1	12.2	13.3	13.7	15.6	21.2	26.9	29.4	Ojikutu (1965)

[a] The description of the population sample is directly taken from that used by the original author(s).

[b] The original paper also gives a breakdown of skin reflectances according to age categories.

[c] A range is quoted here because varying numbers of subjects were tested on the different filters.

[d] The standard deviations of the reflectances at filters 601 to 609 respectively are cited below for the following population samples: Dutch (males): 4.0, 4.4, 4.3, 4.1, 3.9, 3.6, 3.4, 2.6 and 2.5; Irish (males and females): 3.5, 3.6, 3.6, 3.4, 3.1, 3.0, 2.8, 2.7 and 2.7; English (females): 3.5, 3.9, 3.9, 3.8, 3.8, 3.6, 3.0, 2.7 and 2.3; North Indians (males): 4.1, 5.0, 4.6, 5.1, 5.3, 6.1, 5.8, 5.1 and 4.1.

[e] The reflectance readings on this population were taken not on the inner upper arm but on the flexor surface of the forearm at the junction of the middle and proximal thirds.

Genetic studies

The accuracy and objectivity of reflectance spectrophotometry in quantifying skin colour have made it a reliable method for determining zygosity in twins (Collins, Lerner & McGuire, 1966). Indeed, most of the studies that have investigated the genetics of skin colour (see pp. 22–4) obtained their data by reflectance readings. Moreover, as human pigmentation appears to be controlled by a relatively small number of major genes, reflectance spectrophotometry has also been adapted to estimating the amount of admixture in hybrid populations. The success of this approach is apparent from the close agreement between admixture estimates based on skin reflectance data and those derived from blood groups and immunoglobulins (Relethford & Lees, 1981).

Notes

1 It is important to recognize that the naked eye can make fine discriminations of skin colour on direct comparison, and can also take into account broader areas of skin and local variations in colour and shape.
2 Carotenaemia is due to high serum carotene levels that follow the over-ingestion of carotene-containing foods (e.g. carrots, tomatoes). It is endemic in West Africa where red palm oil is used liberally in cooking.
3 The notion that the inner upper arm is unaffected by tanning is not borne out by a number of studies (see p. 112).
4 A very recent study has demonstrated for the first time in humans that UV light induces an increase in the melanocyte population in non-exposed skin areas of the order of 30 per cent of that in the exposed areas (Stierner *et al.*, 1989). *Journal of Investigative Dermatology*, **92**, 561–4). For a possible mechanism of UV-induced melanogenesis in non-exposed areas see note 3, Chapter 3.

8 *Disorders of hyperpigmentation*

Chapter 1 reviewed the biology of human skin pigmentation, and it is now appropriate to consider the pathology of the melanin pigmentary system and, if possible, to clarify the mechanisms producing these abnormalities. Basically, disturbances in human pigmentation manifest clinically as either excessive pigmentation (hyperpigmentation) or deficient pigmentation (hypopigmentation). Any respectable textbook of dermatology will provide lists of the legion conditions which fall under the rubric of the hyperpigmentation and hypopigmentation disorders respectively. Most of these are rare and of no interest to the general reader. This chapter will discuss some selected examples of the hyperpigmentation disorders and the following chapter will consider certain conditions associated with hypopigmentation.

It must be emphasized at the outset that the diagnosis of hyperpigmentation may be difficult. The *normal* skin colour of a Caucasoid of Mediterranean origin, for example, may not differ in the intensity of its hue from the *abnormal* pigmentation of a fair-skinned Scandinavian patient. Furthermore, pigmentation of the oral mucosa (e.g. gums) is usually pathological in fair-skinned Caucasoids but not in the darker ethnic groups (see p. 76).

Hormonal and metabolic factors

Reference was made in Chapter 2 to hyperpigmentation caused by the sex hormones (oestrogens and progesterone) and particularly to the chloasma induced by pregnancy and oral contraceptive agents (see Fig. 2.2).

The classic pathological condition causing hyperpigmentation is *Addison's disease*. This disease, described by Addison in 1855, is due to a failure of the adrenal glands to produce sufficient quantities of the adrenal hormones (corticosteroids). A common cause (and probably the commonest in Third World countries) is tuberculosis of the adrenal glands. Addison's disease results in weakness, lassitude and low blood pressure. Because of the adrenal insufficiency the pituitary gland puts out excessive amounts of adrenocorticotrophic hormone (ACTH) and beta-lipotropic hormone (beta-LPH). As these hormones have melanocyte-

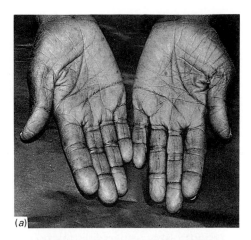

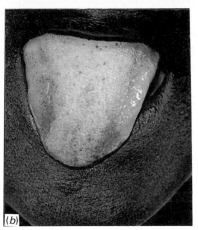

Fig. 8.1(*a*) and (*b*). Patient with Addison's disease showing (*a*) pigmentation of palms and palmar creases, (*b*) pigmentation of tongue. (Courtesy of Dr Martin Abrahamson.)

stimulating activity (see pp. 32–4) pigmentary changes are almost invariable in Addison's disease. An exception is the rare case where the disease is due to failure of pituitary secretion: there would not be abnormal pigmentation because of the lack of ACTH and beta-LPH. However, in classic Addison's disease there is diffuse hyperpigmentation which is most intense on the sun-exposed skin, in the body folds and at sites of friction and pressure (Fig. 8.1(*a*)). Pigmented areas such as the nipples and genitalia darken, and a very common feature is pigmentation of the oral mucosa (Fig. 8.1(*b*)). In some cases dark patches of the skin are mingled with areas of vitiligo (i.e. loss of pigment), giving a black-and-white appearance. This sign *inter alia* led Sir Zachary Cope (1964) to diagnose Jane Austen's last illness as Addison's disease. Against a background of weakness and general debility Jane Austen commented in a letter of 1817, four months before she died (aged 41), 'my looks . . . have been bad enough, black and white and every wrong colour'.

The opposite state to Addison's disease is *Cushing's syndrome,* a condition characterized by an abnormally high secretion of corticosteroids from hyperactive adrenal glands. The treatment sometimes necessitates surgical removal of these glands after which, in about one-third of the patients, a pituitary tumour develops. This situation is known as *Nelson's syndrome,*[1] and it may result in such intense skin hyperpigmentation (Fig. 8.2) that the ethnic identity of affected patients may be called into question, often with resultant psychosocial trauma (see p. 183). Nelson's syndrome is due to the secretion by the pituitary tumour of very large

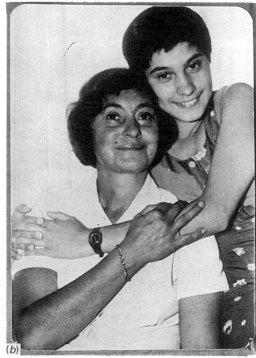

Fig. 8.2(*a*) and (*b*). Caucasoid patient with Nelson's syndrome showing marked hyperpigmentation of skin. (*a*) A family snapshot taken long before condition developed; (*b*) taken at time of her illness shows patient's darker pigmentation against her daughter's skin colour. (Courtesy of Raymond Joseph, *Sunday Times,* Johannesburg.)

quantities of ACTH and beta-LPH, and possibly also to the release of alpha-MSH (which is usually minutely present in human pituitary) (Friedmann & Thody, 1986). Patients may have impairment of vision (even blindness) because of the pressure of the tumour on the adjacent optic nerves. The treatment of the syndrome, either by radiotherapy or pituitary surgery, is very difficult. The anti-epileptic drug, sodium valproate, has been used with some success to suppress the abnormal pituitary secretion and thereby diminish skin pigmentation (Dornhurst *et al.*, 1983). It is alleged that removal of the adrenal glands unmasks a dormant pituitary tumour, and patients with Cushing's syndrome often undergo irradiation of the pituitary gland prior to adrenal surgery. This procedure has prevented the onset of Nelson's syndrome.

Pigmentation of the skin is present in 90 per cent or more of patients with *haemochromatosis,* a disease characterized by grossly excessive iron stores in the body with associated damage to the organs involved. The pigmentation develops very gradually and, although generalized, it is usually most pronounced on the face and sometimes in the oral mucosa. The pigmentation is either a bronze (brown) or slate-grey coloration and it is due to iron pigment and melanin.

Among the porphyrias (see p. 57) there are two cutaneous types (porphyria cutanea tarda and variegate porphyria) which are characterized by increased fragility of the skin in the light-exposed areas. As a result there is blistering, crusting and scarring of the affected parts and in many instances the development of a mottled, chloasma-like hyperpigmentation of the face (Fig. 8.3). These lesions are not associated with acute, painful reactions (as in other photosensitivity disorders) and patients are often not aware that sunlight exposure is responsible for the skin changes, although the condition is worse in the spring and summer. Variegate porphyria is a dominant genetic disorder prevalent among Caucasoid South Africans (usually Afrikaans-speaking). It came into South Africa through two individuals who had been brought out of Holland by the Dutch East India Company and had married in the Cape in 1688. Most of the present-day sufferers of variegate porphyria are descendants of this single family (Dean, 1963). Porphyria cutanea tarda is not a genetic condition but is secondary to factors such as liver disease and toxins (e.g. in alcohol abuse).

Developmental causes
Freckles (ephelides) and lentigines
Freckling is a genetic trait (probably dominant) first appearing during childhood, usually in red-headed or fair-skinned persons. It is

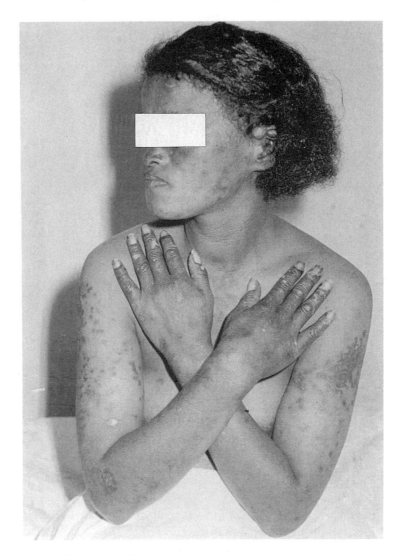

Fig. 8.3 Patient with cutaneous porphyria demonstrating mottled pigmentation of face and upper limbs together with crusting and scarring at these sites. Several scarred areas show depigmentation. (Courtesy of Professor Norma Saxe.)

characterized by the presence of specific groups of melanocytes that have the capacity to form melanin more rapidly after UV exposure than the melanocytes in normal skin. Most freckles are only a millimetre or two in diameter and they increase in number and in pigmentation in summer. There is no increase in the melanocyte density.

The *lentigo* is clinically similar to the freckle although it is darker. It may arise on both exposed and non-exposed sites and it does not have the same clear-cut relationship to UV light as do freckles. Unlike the latter, there is an increased number of melanocytes in the lentigo basal layer. *Senile lentigines* ('liver spots') are common in Caucasoids after prolonged exposure to sunlight, and especially so in the elderly in whom they manifest as slightly roughened areas on the hands and face.

Café-au-lait spots

These uniformly brown spots differ from freckles in being large, more variably shaped and independent of sunlight exposure. They appear at birth (or soon after) on almost any area of skin, either singly or in relatively small numbers. One or two of these lesions are found in 10 per cent of the normal population but the presence of more than five café-au-lait spots with a diameter greater than 1.5 cm points to a diagnosis of *neurofibromatosis* (von Recklinghausen's disease). The latter is a rare genetic condition (autosomal dominant) which develops in late childhood or early adolescence (Fig. 8.4).

On electron microscopy, giant pigment granules (referred to as *macromelanosomes*) are noted in the café-au-lait patches of neurofibromatosis. These macromelanosomes have a diameter of 2.0–5.5 µm (but may occasionally reach 10 µm): they are spherical brown-black bodies situated mainly in the melanocytes and to a lesser extent in the keratinocytes. Neurofibromatosis is characterized by two other signs – axillary freckling and multiple skin or nerve tumours (Fig. 8.4). Occasionally the skin tumours can become so overgrown with subcutaneous tissue that they project hideously from the skin, giving the sufferer a grotesque appearance. This extreme form of the disease was recently publicized in the revival of the case of John Merrick, 'The Elephant Man', who in the 1880s was rescued from the humiliation of being a fairground exhibit and accommodated in a private suite at the London Hospital. Under the care of the eminent surgeon Sir Frederick Treves his innate intelligence and creativity blossomed in the remaining period of his short life.[2]

Mongolian spot

This is an area of blue-black pigmentation, situated over the lower part of the back and the buttocks, which is present at birth (Fig. 8.5). It represents the persistence of melanocytes in the dermis, the blue colour being an optical effect due to Tyndall scattering (see p. 72). The Mongolian spot was so called because it was originally thought to occur exclusively in Mongoloid people, but although it does exist in about 95 per cent of Mongoloid infants, it is also observed in a similar proportion of Negroid babies and in 75 per cent of 'Cape Coloured' newborns. Although it is occasionally seen in dark-complexioned European Caucasoids, it never occurs in blond children with blue eyes. The Mongolian spot usually recedes in the period from 6 months to several years after birth, but in less than 5 per cent of cases it may persist into adulthood. In Mexico during the Colonial Period, when ethnic purity was regarded as an important determinant of social status, the Mongolian spot was deemed to be a stigma of 'mixed blood'. As soon as a child was born, the spot was sought in order to establish if miscegenation had occurred (Ayala Uribe, 1976), and those children with the spot were insulted and called 'purple tails'. Under the South African apartheid system, the

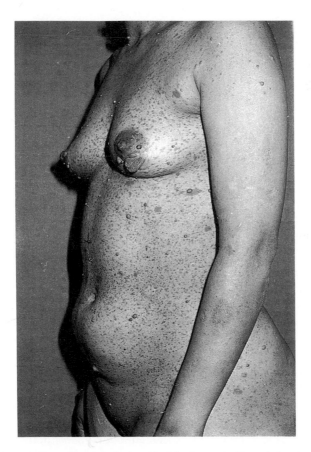

Fig. 8.4 Neurofibromatosis (von Recklinghausen's disease) showing multiple café-au-lait spots and skin tumours. (Courtesy of Professor Peter Beighton.)

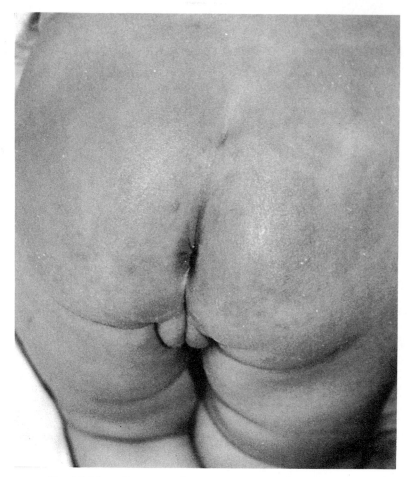

Fig. 8.5 Mongolian spots on buttocks and lower back of 'Cape Coloured' infant. Note extensive areas of (blue-black) discoloration. (Courtesy of Professor Norma Saxe.)

Mongolian spot was one of the criteria for racial classification, especially in deciding between 'white' and 'Cape Coloured' babies.

Pigmented naevi (moles) and melanoma

Naevi are common lesions which increase in number in childhood, reach a peak during puberty and decline in old age. They are aggregates of melanocytes at the epidermal–dermal junction and/or in the dermis. The great majority of naevi are harmless and require no treatment, but certain types of naevi (especially the large variety) can undergo malignant change and transform into *malignant melanomas*. Malignant melanomas can also arise spontaneously from melanocytes and, if they spread to sites distant from the skin, they can be rapidly fatal. There has been concern because since the 1950s the incidence of malignant melanoma has risen steadily by over 5 per cent per annum in fair-skinned Caucasoids all over the world. This increase is partly related to the increasing amount of leisure time spent in the sun, and it seems that it is sunburn caused by repeated (intermittent) high-intensity radiation rather than chronic low-dose exposure that promotes the induction of melanoma (*Lancet*, 1987). In a recent South Africa survey (Rippey & Rippey, 1984), the incidence of malignant melanoma was 2.5 to 6 times more frequent in Caucasoids than in Negroids. It was of interest that, whereas these tumours tended to occur in the sun-exposed areas of Caucasoids, nearly 90 per cent of those in Negroids were found on the soles and palms.

The phenothiazines and schizophrenia

In Chapter 6 the use of the phenothiazines in schizophrenia was discussed. These antipsychotic agents may induce neurological syndromes (e.g. Parkinsonism and tardive dyskinesia) and retinopathy, possibly because of their strong binding to neuromelanin and retinal melanin respectively. However, one of the most intriguing phenomena is the peculiar skin pigmentation caused by phenothiazines.

Greiner & Berry (1964) described a series of 21 female Caucasoid patients from British Columbia, Canada, who exhibited a marked purplish or slate-grey pigmentation (quite unlike suntan) which was mainly limited to the exposed areas of the face, neck and upper chest. They were chronic schizophrenics and all had been receiving the phenothiazine compound, chlorpromazine, in high dosages. Twelve of the patients also showed eye signs – a peculiar hazy-brown pigmentation of the exposed sclera and cornea together with opacity of the lens.

Skin–eye syndrome

This detailed description by Greiner & Berry unleashed a flood of similar reports from all over the world; the condition was termed the *skin–eye syndrome* and the pigmented patients have been dubbed 'the purple people'. The pigmentation is confined to the exposed areas of the body (being accentuated in the summer months); it is unequivocally associated with the prolonged and high-dosage administration of phenothiazines (usually chlorpromazine), and on cessation of the medication it tends to persist. The eye effects are fairly characteristic. They start as a fine, dust-like stippling of yellowish-white to golden-brown opacities in the anterior portion of the lens. These opacities merge with one another until they eventually form an anterior cataract. The corneal changes consist of an accumulation of similar dot-like, brownish-yellow granules which, in severe cases, can infiltrate the entire cornea. These lens and corneal signs are concentrated centrally and usually affect both eyes; they occur in about one-third of phenothiazine-treated patients and their severity is related to the total phenothiazine dosage received. Affected patients suffer no visual impairment, and this is fortunate as many schizophrenics require prolonged phenothiazine therapy for control of the illness.

Visceral melanin

The pigment involved in the skin hyperpigmentation appears to be melanin, and the purple or slate-grey colour is explained (on the basis of the Tyndall scattering effect) by the presence of melanosomes in the dermis. These dermal melanosomes are not located in melanocytes (which are actually absent from the dermis) but in phagocytic cells, and the appearance of the latter may also explain the surprising pigmentation observed in internal organs by Greiner & Nicolson (1964). They did autopsies on 12 patients with skin pigmentation who had died unexpectedly and noted diffuse deposition of pigment (which they identified as melanin) in liver, heart, kidney, lungs, brain and other organs. Their findings are controversial and corroborative studies are required. Even if visceral pigment was present in these patients it is much more likely to have been a substance other than melanin (e.g. phenothiazine itself or a metabolite). Wassermann (1974), however, believed that this visceral pigmentation was compatible with the leucocytic transport of melanin (see p. 88). He demonstrated by the skin-window technique that the topical application of chlorpromazine in Caucasoid skin led to the appearance first of blue material (chlorpromazine) in leucocytes and then of melanosomes. He formed the hypothesis that the leucocytic chlorpromazine (being an electron donor) attracted the electron acceptor, melanin, and became bound to it; this process triggered the leucocyte transport system which then distributed melanin to the various organs.

Mechanism of phenothiazine skin pigmentation

The prevalence of skin hyperpigmentation among chronic phenothiazine-treated schizophrenics is approximately 1 per cent on average, and most surveys have found a marked preponderance of females. This relatively rare occurrence suggests that the drug reaction is idiosyncratic (rather than being common and expected like phenothiazine-induced Parkinsonism). Bolt & Forrest (1968) proposed that chlorpromazine pigmentation represented an interaction between a metabolite of chlorpromazine (7-hydroxychlorpromazine) and melanoprotein, in the presence of light, to form a complete charge-transfer reaction. They also speculated that the 'purple people' had a genetic metabolic error in the detoxification of the drug. A recent microanalytical study of the disorder has strongly suggested that pigmentation is due, at least in part, to the presence in the dermis of dense inclusion bodies containing chlorpromazine or its metabolite (Benning *et al.*, 1988). However, even patients on long-term phenothiazine (chlorpromazine) therapy who had no sign of clinically abnormal pigmentation did show increased skin melanin concentrations on reflectance spectrophotometry (Robins, 1975). These patients (Fig. 8.6) consisted of South African Negroids and Caucasoids, and it was interesting that in both ethnic groups females had more marked hyperpigmentation than males. Moreover, the Caucasoids (but not the Negroids) had a greater increase of melanin in the exposed areas.

A possible hormonal factor in phenothiazine pigmentation has been investigated. Although phenothiazines increased the secretion of MSH from the pituitary glands of experimental animals, these drugs (administered long-term or short-term) had no effect on the levels of 'plasma immunoreactive beta-MSH' in patients, although some of them manifested clinical phenothiazine pigmentation (Smith *et al.*, 1977a). As beta-MSH does not occur in the human (see p. 33) the compound measured in the latter study was probably beta-LPH (or a breakdown product), a hormone that is evidently not affected by phenothiazines. It is also possible that phenothiazines have a direct effect on epidermal melanocytes, especially as they activate the tyrosinase in melanoma tissue in proportion to their antipsychotic properties (Van Woert, 1970). Whatever the mechanism, it appears that female sex hormones and UV exposure potentiate the increased melanogenesis due to phenothiazines.

Mechanism of lens and corneal pigmentation
The purplish skin pigmentation is rare among phenothiazine-treated patients, but corneal and lens changes occur in about one-third of all patients. The nature of the eye deposits is unknown, but the distribution of the opacities in the central area of the lens and cornea suggests that photosensitivity (see p. 55) is a likely factor. The phenothiazines are well-established photosensitizing agents, and chlorpromazine is the most potent photosensitizer within the phenothiazine group. It seems that the patients on high-dosage phenothiazine regimens who sunburn easily are more susceptible to eye changes than less photosensitive individuals (De Long, 1968).

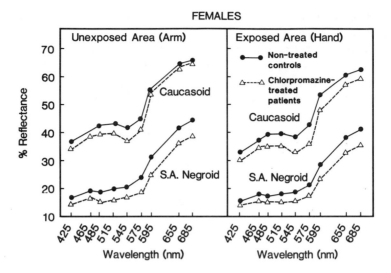

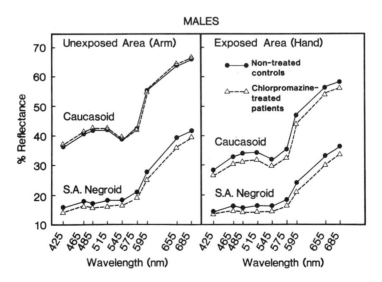

Fig. 8.6 Skin reflectance curves (EEL Reflectance Spectrophotometer) in female and male groups of chlorpromazine-treated Caucasoid and South African Negroid patients, compared with non-treated controls. In general, chlorpromazine-treated patients have a marked increase in skin pigmentation (i.e. lower skin reflectances).

The skin component of the skin–eye syndrome is not due to photosensitivity, since microscopic examination of the pigmented skin shows none of the inflammatory changes typical of sunburn. However, the eye opacities may be the result of a light-induced interaction of the photosensitizing phenothiazine with tissue protein, or they may actually represent deposits of the drug within the lens. Ocular melanin does not seem to play any part in this process because chlorpromazine-treated guinea-pigs developed cataracts whether they were pigmented or albino (Howard *et al.*, 1969).

It had been suggested that the skin and eye changes described might be external signs of possible brain damage caused by the phenothiazines. However, no association whatsoever was found between phenothiazine-induced Parkinsonism or tardive dyskinesia (see pp. 89–90) and either skin or ocular pigmentation (Wheeler, Bhalerao & Gilkes, 1969; Ananth & Yassa, 1982).

Schizophrenia melanosis

Greiner & Nicolson (1965) put forward the concept of 'schizophrenia melanosis'. Essentially they questioned whether the phenothiazine pigmentation in schizophrenic patients was primarily due to the schizophrenia or to the phenothiazines. They accordingly re-examined autopsy material from schizophrenic patients who had died between 1947 and 1949, several years prior to the introduction of phenothiazines. Melanin was present in a similar distribution to that described in phenothiazine-treated patients (Greiner & Nicolson, 1964), but in lesser amounts. They developed the theory that melanogenesis was already increased in schizophrenia and that phenothiazine therapy merely augmented it. Their hypothesis implied that schizophrenia was genetically linked to a biochemical abnormality that produced both the psychotic symptoms and the hyperpigmentation.[3]

The occurrence of hyperpigmentation in schizophrenic patients was not a new phenomenon. Earlier in the twentieth century several authors had commented on the association of abnormal pigmentation with schizophrenia (Derbes, Fleming & Becker, 1955), one of the first being Wigert who in 1926 described a striking bronze pigmentation (*Bronzkatatonie*) in four patients with catatonic schizophrenia. Whereas many scientists might dismiss this association as 'anecdotal' or 'scientifically improbable', it does have a theoretical basis. The melanocyte and the neurone ultimately share the same developmental origin in the neural plate, and both cell types use tyrosine as a precursor in the biosynthesis of their specific products (melanin and the catecholamines respectively).

Recent studies using objective measurements of skin colour (e.g. reflectance spectrophotometry) and including control groups (e.g. non-schizophrenic psychiatric patients) have failed to confirm that schizophrenics untreated with phenothiazine medication have increased skin melanin concentrations (Robins, 1972; Wassermann, 1974; Lodge Patch, 1975). Thus, the exciting contention by Greiner and Nicolson of a schizophrenia melanosis has been laid to rest – 'The tragedy of science', as T. H. Huxley put it, is 'the slaying of a beautiful hypothesis by a single fact'. But what of those early reports of increased pigmentation in schizophrenia? The schizophrenic patients undoubtedly had abnormal pigmentation – but it was due not to the schizophrenic illness but to environmental and nutritional factors. In the days before antipsychotic medicines schizophrenic patients were often kept in custodial care, usually in the chronic wards of mental hospitals where they were subject to poor hygiene, infection and dietary insufficiency. The last may have been important because vitamins A and C and several of the B group vitamins can influence the process of melanogenesis. (Deficiency of nicotinic acid, a member of the B group, causes pellagra, of which a notable feature is hyperpigmentation of the exposed skin areas.) These unforeseen pitfalls have persistently dogged research into schizophrenia and ensnared many an unwary investigator. The truth of the matter is that, from the standpoint of understanding the schizophrenic process itself, bronzed catatonics and purple people are red herrings!

Other agents causing hyperpigmentation

Certain anti-epileptic agents belonging to the hydantoin group, of which phenytoin is the best known, may cause skin hyperpigmentation (Levantine & Almeyda, 1973). The long-term use of the hydantoins produces a brown or bronze melanin pigmentation on the face and neck, akin to the chloasma of pregnancy and with a definite female preponderance (Kuske & Krebs, 1964). Skin reflectance spectrophotometry on a random sample of phenytoin-treated patients without overt hyperpigmentation has revealed increased melanin concentrations, particularly in females and at the exposed body areas of Caucasoids (Robins, 1972). This pattern closely resembled that of the phenothiazine pigmentation described above.

The affinity of chloroquine for melanin and the potential retinotoxicity of this drug have been discussed (see p. 91). Surprisingly, the most conspicuous effect of chloroquine is a depigmentation (bleaching) of the hair – not confined to the scalp but affecting the hairs everywhere, including the armpit and pubic area (Dupré *et al.*, 1985). This bleaching is reversible on drug withdrawal; it only affects redheads and blonds (whose hair becomes white) and it is without effect on black hair. Skin is virtually never involved in the bleaching phenomenon: on the contrary, chloroquine (especially if taken on a long-term basis) is a well-documented cause of skin hyperpigmentation (Levantine & Almeyda, 1973). It usually manifests as a grey-to-blue-black discoloration of the skin and oral mucosa which is probably due to the combination of a melanin–chloroquine complex with an iron pigment. Melanin is present in the dermis of patients with chloroquine-induced pigmentation (Levy, 1982).

Prior to the advent of chloroquine in the treatment of malaria, mepacrine (quinacrine) was a commonly used agent. Among its side-effects, mepacrine caused a diffuse yellow skin pigmentation affecting the limbs, forehead, face and trunk. During the Second World War this discoloration was a regular sight among troops in the malarial areas. The pigmentation appeared within 3–10 days after the start of mepacrine use and persisted for from a few weeks to several months after its cessation. The yellow pigmentation was attributed to the yellow colour of the drug, which was highly concentrated in the epidermis.

The tetracycline antibiotic, minocycline, after long-term and high-dosage administration in the treatment of acne, causes a blue or blue-black hyperpigmentation predominantly affecting the lower extremities and the acne scar tissue. This discoloration is not due to melanin, but to the accumulation of iron-containing particles within the dermis (Sato *et al.*, 1981).

An interesting agent in relation to skin pigmentation is levodopa, which is the treatment of choice for Parkinson's disease (see p. 82). Because dopa is metabolized to melanin within the melanosome, dopa therapy (given as levodopa) may possibly enhance melanin biosynthesis, although the literature makes scant reference to this point. It was reported that a 51-year-old man with Parkinson's disease developed repigmentation of his white beard after 8 months of levodopa therapy (Grainger, 1973). A pilot trial found that levodopa combined with UV light produced repigmentation of the depigmented areas in four out of seven women with vitiligo (Goolamali, 1973). A small controlled study showed that levodopa was associated with a (non-significant) trend towards darkening of the exposed areas in female patients with Parkinson's disease but not in the corresponding areas of levodopa-treated males (Robins, 1979). There has also been a suspicion that levodopa may promote the growth of malignant melanoma, but evidence for this is unconvincing. Thus, contrary to theoretical expectations, levodopa has minimal effects (if any) on the skin pigmentation of treated patients.

Heavy metals have been implicated in cases of melanin hyperpigmentation which arose after the excessive use of medicines containing arsenic, bismuth, gold or silver (Molokhia & Portnoy, 1973). The metals are believed to act by binding, and thereby inactivating, the sulphydryl compounds in skin which normally inhibit tyrosinase activity (see p. 30). The removal of this inhibition stimulates melanogenesis. Mercury products inactivate tyrosinase (probably by replacing copper in that enzyme) and they have been used for cosmetic purposes as skin bleaching agents. However, in prolonged use mercury may also bind to sulphydryl groups with the opposite effect of hyperpigmentation.

Notes

1 Nowadays removal of the adrenal glands is avoided in the management of most patients with Cushing's syndrome. Recent technical advances have made pituitary surgery the preferred treatment for the disorder and this has reduced the emergence of Nelson's syndrome.
2 The diagnosis of neurofibromatosis in the 'Elephant Man' has been challenged by Tibbles & Cohen (1986, *British Medical Journal*, **293**, 683–5).
3 The hypothesis also implies that albinos are protected from the development of schizophrenia. This suggestion, however, is not substantiated because of the occurrence of cases of coexistent schizophrenia and oculocutaneous albinism (Robins, 1980; Pollack & Manschreck, 1986) (see also p. 147).
4 Experiments with mice and guinea-pigs showed that synthetic phosphorylated levodopa, but not levodopa itself, increases skin pigmentation (Bolognia, J. *et al.* (1989). *Journal of Investigative Dermatology*, **92**, 651–6).

9   *Disorders of hypopigmentation*

Albinism

Albinism consists of a group of genetic disorders of the melanin pigmentary system which occurs throughout the animal kingdom from insects, fish and birds right up to human beings. It is characterized by an absence of or decrease in melanin which, in the human varieties of albinism, takes two forms: *oculocutaneous albinism* and *ocular albinism*. The former (which is by far the commoner) manifests as a lack of pigmentation in the skin, hair and eyes; in ocular albinism the loss of melanin is limited to the eyes and skin pigmentation is normal. All human albinos have visual problems – there is hypopigmentation of the iris, choroid and retina as well as maldevelopment of the fovea, a part of the retina which mediates central vision. The typical eye signs are photophobia (an abnormal, often painful, sensitivity to sunlight leading to its avoidance), nystagmus (involuntary, rhythmical oscillations of the eyeballs, usually in a horizontal plane), squint and a decreased visual acuity (in severe cases amounting to partial blindness). This chapter will concern itself only with oculocutaneous albinism.

History

Allusions to albinism date from antiquity but the actual term 'albino' (from the Latin *albus*, white) was coined by the seventeenth-century Portuguese explorer, Balthazer Tellez, who sighted certain 'white' Negroids on the west coast of Africa. Columbus, however, was claimed to have encountered such people (near Trinidad) at the time of his fourth voyage to America in 1502. The identification of albinos was hardly a feat of recognition: compared with normally pigmented Negroids, these albinos were highly conspicuous, and it was noted that their marked photophobia confined them to their huts until twilight. Their intense visual sensitivity to sunlight led the eighteenth-century Swedish naturalist Linnaeus to describe albinos as '*Homo nocturnus*' and to group them with cave-dwellers.

The legend of the existence of a 'white native race' had been propagated from the earliest times in Greek and Roman writings, and it was only in the nineteenth century that the disorder was viewed more scientifically – an approach that culminated in the realization that albinism represented an inherited biochemical abnormality. Garrod (1908) classified albinism as an inborn error of metabolism due to lack of the melanin-synthesizing enzyme.

Prevalence

The estimated frequency of oculocutaneous albinism varies among different ethnic groups and in different geographical areas. Among European and North American Caucasoids the frequency usually lies between 1:10 000 and 1:20 000 and among North American Negroids it is about 1:10 000. Albinism appears to be more common in African Negroids, being about 1:5000 in Nigeria and 1:3900 in South Africa (Kromberg & Jenkins, 1982). In South Africa, consanguinity appears to be a significant factor, the rate of consanguineous matings, for example, being very high (42 per cent) among the parents of Tswana albinos (Kromberg & Jenkins, 1982). (The Tswana social system encourages marriages between cousins.) There is considerable variation between the different groups of South African Negroids – the Southern Sotho having prevalence rates as high as 1:2000, the Zulu and Xhosa 1:4500 and the Shangaan people 1:29 000. There is a particularly high prevalence of albinism (about 0.7 per cent) among the Cuna Amerindians of Panama and particularly among the 30 000 who live on the San Blas islands, which skirt the Caribbean coast of Panama. The Amerindian groups living in the southwestern United States of America (Zuni, Hopi and Jemez) also have a very high prevalence (about 0.5 per cent). The highest recorded frequency of albinism in the world occurs in the Brandywine triracial isolate of Maryland where 1.2 per cent of the total population are affected (Witkop *et al.*, 1972), all of these albinos deriving from a single pair of ancestors who were among the founders of the present population.

Classification of oculocutaneous albinism

Sir Archibald Garrod's (1908) delineation of albinism as a genetic metabolic disease was a major advance in its history. Since then there has been an unremitting fragmentation of the condition so that, from being conceived as a simple unitary disorder in 1908, it is presently divided into ten different types (Witkop, Quevedo & Fitzpatrick, 1983). The discussion that follows will be limited to the two most important types, namely, *tyrosinase-negative oculocutaneous albinism* (tyr-neg) and *tyrosinase-positive oculocutaneous albinism* (tyr-pos).

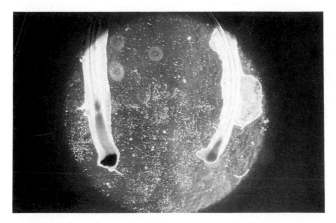

Fig. 9.1 Hair bulb incubation test in oculocutaneous albinism. Hair bulb on left incubated in tyrosine solution, the other in buffer (control) solution. Microscopic examination 12 to 15 hours later showed development of pigment in tyrosine-incubated hair bulb, thus establishing diagnosis of tyrosinase-positive albinism. (No hair bulb pigment forms after tyrosine incubation if tyrosinase-negative.) (Courtesy of Dr Jennifer Kromberg.)

Differentiation between these two types of albinism rests on a fairly simple test. Hairs are plucked from patients and the hair bulbs are incubated in a solution of tyrosine or dopa; 12–15 hours later the bulbs are observed by microscopy to determine whether or not pigment formation has occurred (Fig. 9.1). If there is no discernible pigment under microscopy, then the albinism is classified as tyr-neg; if pigment granules are visible, then it is classed as tyr-pos. Thus, the earlier notion that albinism is due to a lack of the enzyme tyrosinase is not strictly accurate.

Genetics

Both important types of albinism are transmitted by Mendelian recessive inheritance. The albino must be homozygous for the albinism gene, i.e. must have received it from both parents. Individuals who possess only one gene (heterozygote state) are not albino. The clinically normal parents of an albino are heterozygous carriers of the albinism gene, and, according to Mendelian genetics, the mating of two such heterozygote carriers will result in a 1 in 4 probability that an albino baby will be born. If both parents are albinos then all of the offspring will be albino. Tyr-neg albinism is determined by a different recessive gene from that of tyr-pos albinism. This has been convincingly demonstrated in those exceptional cases where normally pigmented children were born to parents, both of whom were albino. On subsequent hair bulb incubation it was found that the parents had different types of albinism.

Tyrosinase-negative (tyr-neg) albinism

This is the classical type of oculocutaneous albinism. There is no clinically detectable pigment in the skin, hair or eyes, and no pigment is observed on hair bulb incubation. It must be emphasized that albinism is not due to any structural abnormality of melanocytes. Moreover, melanocytes are present in their usual numbers in these tissues: the defect resides in the melanization of melanosomes. On electron microscopy only immature stage 1 and 2 melanosomes are seen (see p. 9); tyrosinase is not available for melanin formation and thus the melanosomes cannot progress to stages 3 and 4. All tyr-neg albinos have snow-white hair, pink-white skin, and grey-to-blue-grey irises. The latter are so translucent when transilluminated that the lens of the eye can be seen. There are moderate-to-severe signs of visual impairment (Fig. 9.2).

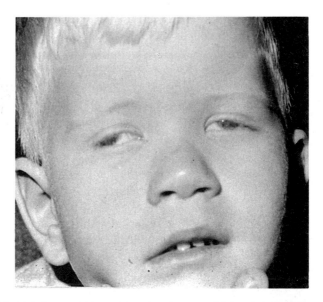

Fig. 9.2 Caucasoid boy with oculocutaneous albinism (tyrosinase-negative) showing lack of melanin in skin, hair and eyes. Irises are blue-grey (translucent on transillumination). Evidence of a squint; patient had marked nystagmus and photophobia and was practically blind, visual acuity being 5 per cent of normal. Examination of fundus revealed absence of pigmentation. (Courtesy of Professor Justin van Selm.)

✓*Tyrosinase-positive (tyr-pos) albinism*

These albinos have detectable pigment in the skin, hair and eyes (Fig. 9.3). The hair bulb test (see Fig. 9.1) shows signs of pigment (and therefore the presence of tyrosinase activity) and on electron microscopy stage 3 (but not stage 4) melanosomes are identified. Although in infancy it is difficult clinically to distinguish between a tyr-neg and a tyr-pos albino, the aging process leads to a gradual accumulation of pigment in the latter so that differentiation then becomes possible. In the older tyr-pos albino, for example, coloration of the hair and eyes has changed to a yellowish or yellowish-brown hue and skin colour is less pinkish-red. The irises are less translucent on transillumination; visual impairment is milder than in the tyr-neg patient and it may even improve with age.

From an epidemiological standpoint, tyr-pos albinism is far commoner in Negroids than is the tyr-neg variety. In Negroid Americans the tyr-pos condition is twice as common (Witkop *et al.*, 1972), and among a sample of South African Negroids it was by far the dominant type (Kromberg, 1985). A survey of Nigerian albinos revealed no tyr-neg subjects (King *et*

Fig. 9.3 South African Negroid family in which father and baby have tyrosinase-positive oculocutaneous albinism. Note discernible pigmentation of father's beard, a feature compatible with tyrosinase-positive albinism. (Courtesy of Dr Jennifer Kromberg.)

al., 1980); among Caucasoids there is an approximately equal prevalence of tyr-neg and tyr-pos albinism, although some studies suggest a preponderance of the latter.

In essence, tyr-neg albinism is associated with a lack of tyrosinase activity but in tyr-pos albinism there is unequivocal evidence of tyrosinase activity (Witkop *et al.*, 1983). The basic biochemical defect in the tyr-pos disorder is unknown. The newly discovered enzyme (Barber *et al.*, 1984) in the tyrosine–melanin biosynthetic pathway, dopachrome oxidoreductase, is alleged to block melanin formation at the level of 5,6-dihydroxyindole (see p. 29). This enzyme possibly plays an active regulatory role in melanogenesis, and it may be implicated in tyr-pos albinism.

The other types of oculocutaneous albinism include the yellow-mutant variety (with hair that can vary from yellow to golden), brown albinism and rufous albinism. The last two are found in Negroids and New Guineans, and they are characterized by skin and hair colours which are brownish and mahogany-like reddish-brown respectively. Although the classification of albinism may appear clear-cut, it is sometimes difficult in practice to assign albinos to a watertight category, even with the aid of techniques such as hair bulb incubation. The development of specific biochemical tests is awaited, and one lead has been the report that a tyr-neg albino had a different urinary excretion pattern of melanin precursors and metabolites to two tyr-pos albinos (Carstam *et al.*, 1985).

Skin cancer and life expectancy

The loss of skin melanin renders albinos (particularly those in tropical and subtropical climes) exceedingly liable to skin cancers (Fig. 9.4). For example, skin cancers in South African Negroid albinos showed a thousandfold and a tenfold higher incidence than in normally pigmented South African Negroids and American Caucasoids respectively, being present even during childhood and adolescence and in all those aged 50 years and over (Kromberg, 1985; Kromberg *et al.*, 1989). The situation is worse nearer the equator where UV intensity is greater. In Nigeria and Tanzania no albino over the age of 20 years was found to be free of malignant or pre-malignant skin lesions (Okoro, 1975; Luande, Hensche & Mohammed, 1985), and in Tanzania chronic skin damage was evident in every albino by the first year of life. The albinos in all three surveys died prematurely from skin cancers, life expectancy being markedly reduced in those from Nigeria and Tanzania where less than 10 per cent survived beyond the age of 30. In South Africa the decreased life span of albinos applied only to the males, possibly because they had greater occupational

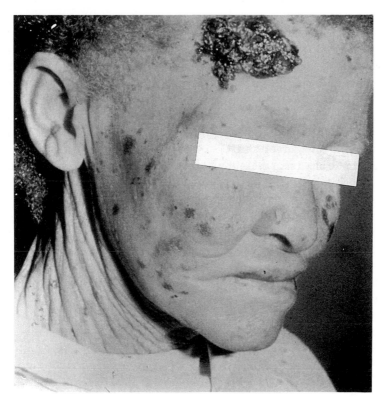

Fig. 9.4 South African ('Cape Coloured') albino woman with large ulcerating skin cancer (squamous cell carcinoma) on forehead. Wrinkled appearance of neck is evidence of premature skin ageing (she is under 40 years of age) from sun-induced damage. (Courtesy of Professor Norma Saxe.)

exposure to the sun (Kromberg, 1985). It is of interest that malignant melanoma is rare in Negroid albinos and only one case of it was encountered in the studies cited above.

Abnormal nerve tracts

Neuromelanin is present in normal amounts in the brains of albino patients (Foley & Baxter, 1958; Marsden, 1965). However, there are structural defects in the albino brain which involve the optical system. Albino animals and humans (with any type of albinism) have aberrant neuronal pathways – there is an abnormal increase in the crossing of optic nerve fibres from the retina to the opposite side of the brain (Witkop *et al.*, 1983) with a marked decrease (or even absence) of ipsilateral fibres. This anatomical flaw causes asymmetry within the cerebral hemispheres, with resultant loss of binocular vision and depth perception; and it would explain such visual disturbances as the nystagmus and squint. This asymmetry is also reflected in the abnormal brain wave recordings (visual evoked potentials) obtained from human albinos (King *et al.*, 1980; Apkarian *et al.*, 1983). There is the hypothesis that embryologically the pigment destined for the retina is somehow necessary for the correct wiring of the optic neurones to their specific brain targets.

Melanin is absent from the inner ear in albino animals (see p. 77), and recent work has demonstrated decreased neuronal size and dendritic density in the central auditory brain structures of albino animals (Conlee, Parks & Creel, 1986b). An evaluation of brainstem auditory evoked potentials in human albinos revealed hemispheric asymmetries that paralleled those in the visual system (Creel *et al.*, 1980; Garber *et al.*, 1982). The clinical significance of these electrophysiological findings is not yet clear, although in guinea-pig experiments the albino animals were more susceptible to noise-induced trauma, a possible effect of absent cochlear melanin (Conlee *et al.*, 1986b).

Intelligence and personality

Stewart & Keeler (1965) did a psychological study of six albino and six non-albino San Blas Cuna Amerindians. The two groups did not differ in intelligence, but in personality the albinos were less stable, less aggressive, more immature, and less communicative. Beckham (1946) investigated 42 Negroid albino schoolchildren and found that they did not differ in intelligence from their non-albino siblings but had a general feeling of insecurity which he attributed to their sense of being different. In a comparison of 28 South African Negroid albinos with 28 matched normal controls the albinos were found to be slightly more intellectually mature but inclined towards a negative self-evaluation (Manganyi, Kromberg & Jenkins, 1974). Psychometric testing of 12 albino children revealed that their verbal IQs were strikingly higher than their performance IQs, a pattern not observed in 12 non-albino control children with equal visual handicap (Cole *et al.*, 1987). This phenomenon may relate to the anomalous visual pathways and association areas in albinism. In general, albinos perform intellectually, socially and occupationally as well as – if not better than – their non-albino peers and may compensate for their visual handicaps by more dedicated intellectual activity and by the superior development of memory (Taylor, 1987).[1] If they have minor personality and emotional difficulties then these are probably due mainly to their interaction with society, which in several countries views albinos with hostility and prejudice (see pp. 180–1).

l Neuropsychiatric effects

Epilepsy is rarely associated with albinism. The ocular hypopigmentation and nystagmus of albinos have raised the suspicion that such patients might be more sensitive to the form of epilepsy initiated by flickering light (photogenic epilepsy), but an electroencephalographic study of South African Negroid albinos during intermittent photic stimulation (light flickering) showed no evidence of a photogenic trait (Bental, 1979). A preliminary study (Oosthuizen *et al.*, 1983) suggests that albinos have raised blood levels of melatonin throughout the day and night, with loss of the normal rhythm in which melatonin levels peaked at around midnight (see p. 35). If confirmed, this finding may relate to the deranged optic tracts in albinism because melatonin secretion is controlled by relays of nerve fibres originating in the retina. Finally, the sporadic reports of coexistent albinism and schizophrenia are probably best explained on the basis of a chance occurrence of the two conditions in the same individual (Robins, 1980; Pollack & Manschreck, 1986), although the very rare familial association of the two disorders suggests either genetic linkage or a neurochemical (neuroanatomical) causal connection (Clarke & Buckley, 1989).

Inappropriate use of albino animals in research

It has become common in animal research to use albino animals, particularly rats and mice. There is no logical basis for this practice and it arose accidentally with albino strains of rats being perpetuated for show purposes during the nineteenth century. These strains were then introduced into laboratories, such as the Wistar Institute from which the famous 'Wistar strain' was bred. The fashion for albino experimental animals developed from two motives – the first that these animals were considered to be aesthetically pleasing; the second (and stronger one) that albinos were deemed to be more docile and therefore more manageable in the laboratory. The idea that non-aggression is dependent on albinism *per se* is erroneous (Creel, 1980): black rats from certain strains can be non-aggressive and albino rats from other strains can be vicious.

The inappropriateness of using albinos in research has been cogently argued by Creel (1980). The optical and auditory abnormalities associated with albinism must undoubtedly affect the behaviour of the animals, especially in psychological experiments involving visuo-spatial and auditory learning tasks. Another area where testing of albino animals may generate misleading data is in the field of drug evaluation. Certain drugs (e.g. chloroquine, phenothiazines) have a strong affinity for melanin (see pp. 89–92) and this may account for their potential to produce retinal and inner ear toxicity (both tissues containing melanin). In albino animals a drug like chloroquine does not induce retinotoxicity, and the results of drug studies on albinos may not necessarily apply to the normal, pigmented animals. Certainly, in the case of a drug with a potential for blindness (retinotoxicity) or deafness (ototoxicity), false conclusions may be derived about its safety if it is investigated only in albino experimental animals.

Management of the albino patient

Albino patients must take strict precautions against sun exposure. Whenever possible, the sun should be avoided, particularly between 10.00 and 14.00 in summer. In addition to protective clothing the use of the physical barrier creams (e.g. those containing titanium dioxide or zinc oxide), rather than the conventional sunscreen agents, is desirable. Regular inspection of the skin (at least yearly) is indicated to search for incipient malignant changes. Examination of the eyes will be required for the prescription of correct spectacles, and tinted glasses may help to reduce photophobia. The psychosocial aspects of the condition deserve special attention – albino children *and* their parents require specific counselling about the problems of albinism and the relevant management stategies. In some societies ignorance about their condition has gravely aggravated the plight of albinos, making them victims instead of the recipients of care and understanding. The British Albino Fellowship (Taylor, 1980) is an excellent example of a welfare organization for albinos.

Genetic counselling

As oculocutaneous albinism is a Mendelian recessive disorder, each parent of an albino child is a heterozygote for the gene and therefore a carrier. The mating of two heterozygotes (carriers) will result in a 25 per cent probability of producing an albino child (who is homozygous for the gene), a 50 per cent probability of producing a heterozygote carrier, and a 25 per cent probability of a normal child. If the frequency of albinism in a given population is, say, 1 in 10 000, then the carrier rate is approximately 1 in 50; if the frequency is 1 in 3600, then the carrier rate is approximately 1 in 30.

Genetic counselling means imparting this type of information to those parents and prospective parents with a family history of albinism. The normally pigmented sibling of an albino, for instance, will want to know what are his or her chances of having an albino child.

If both parents are non-albino, then he or she has a 2 in 3 chance of being a heterozygote carrier and a 1 in 3 chance of being totally free of the albinism gene. But instead of this kind of statistical advice, it would be much more useful to establish definitely whether that person is, or is not, a heterozygote carrier of the albinism gene.

The feasibility of determining heterozygosity has been thoroughly investigated and there are a few pointers. Among some Caucasoid (but not Negroid) heterozygotes for tyr-neg albinism, there is a trend for the iris to show abnormal translucency on transillumination. This sign, however, is not specific enough to make it a reliable index and it also fails to select out the tyr-pos carriers. A test that has proved to be more valuable is the hair bulb tyrosinase assay (King & Olds, 1985). In tyr-neg heterozygotes tyrosinase activity (as measured by the assay) is very low (often zero) and the test has some validity; but with tyr-pos heterozygotes the assay results are so variable that the test becomes unacceptable.

Another approach to studying carriers was to determine whether their skin pigmentation is decreased. Reflectance spectrophotometry showed that South African Negroids who were heterozygotes for albinism (presumed to be tyr-pos although not tested) had significantly lighter skin colour (higher reflectances) than unaffected control subjects (Kromberg, 1985; Roberts, Kromberg & Jenkins, 1986) (Figs. 9.5 and 9.6). The differences between carriers and non-carrier controls were not sufficiently discriminating for the procedure to be recommended as a diagnostic tool.

The optical anomalies of albinism have led to the speculation that carriers of the gene might manifest subtle visual disturbances on electrophysiological testing. Visual evoked potentials have been measured in South African Negroid heterozygotes (tyr-pos), but no specific aberrations were detected (Castle *et al.*, 1988) – thereby eliminating this method as a detection test.

Finally, there is the question of the prenatal diagnosis of oculocutaneous albinism. This was successfully achieved in the case of a 36-year-old Middle Eastern woman whose first child was an albino (type unknown) (Eady *et al.*, 1983). At 20 weeks of gestation foetoscopy was performed and skin samples were taken from the foetal scalp under direct vision. Electron microscopy of the tissue revealed that melanosomes in the hair bulb melanocytes had progressed no further than stage 2, reflecting a lack of melanin synthesis. In four control foetuses of the same gestational age there were numerous stage 4 melanosomes, indicating active melanin formation. The pregnancy was terminated at 22 weeks and the diagnosis of oculocutaneous albinism was confirmed on postmortem examination.

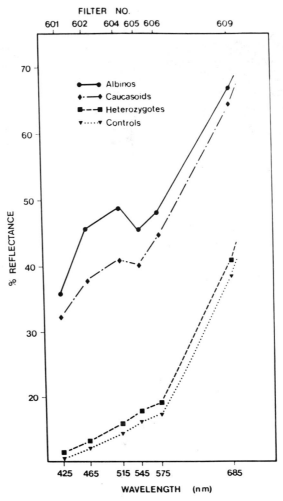

Fig. 9.5 Skin reflectances at six wavelengths (EEL reflectance spectrophotometer) from inner aspect of upper arm in four groups of subjects: Southern African Negroid albinos (male and female), Southern African Caucasoids (female), Southern African Negroid heterozygotes (mothers of albinos) and Southern African Negroid controls (female). Heterozygotes significantly less pigmented than controls and Negroid albinos lighter in skin colour than Caucasoids. (From Kromberg, 1985, with permission.)

However, although techniques to detect albinism during pregnancy are available, the case for termination has been argued against by Taylor (1987). Certainly in northern regions such as the United Kingdom (where sunshine is not intense enough to generate skin cancer) he contends that

Fig. 9.6 Southern African Negroid heterozygote for oculocutaneous albinism (mother of albino) (left) with markedly lighter skin pigmentation than normally pigmented woman (right). (Courtesy of Dr Jennifer Kromberg.)

there is little justification for abortion, especially as the albino is capable of leading a full and creative life provided that the preventive and management techniques outlined above are adopted.

The recent discovery of the human tyrosinase gene on the long arm of chromosome 11 (q14–q21) (Barton, Kwon & Francke, 1988) and of specific mutations of this gene in tyr-neg albinos (Tomita *et al.*, 1989; Spritz *et al.*, 1990) opens up the possibility not only of the prenatal diagnosis of tyr-neg albinism by amniocentesis but also of carrier detection.

Phenylketonuria

Phenylketonuria is an inborn error of metabolism which, like oculocutaneous albinism, is due to an autosomal recessive gene. It is characterized by mental retardation and by a dilution in the pigmentation of the skin, hair and eyes. Phenylketonuria is due to a deficiency of the enzyme phenylalanine hydroxylase, the function of which is to convert phenylalanine into tyrosine. Tyrosine is the point of departure for two important pathways: the conversion of tyrosine to melanin in the melanocyte (involving tyrosinase) and the conversion of tyrosine to catecholamines (dopamine, adrenaline and noradrenaline) in neuronal tissue and adrenal medulla (involving the enzyme tyrosine hydroxylase).

A deficiency of phenylalanine hydroxylase causes a massive build-up of phenylalanine in the blood, and this in turn has metabolic repercussions which arrest normal brain development. After 6 months of life mental retardation is evident, and thereafter phenylketonuric patients become severely mentally handicapped (IQ 20–25). They have a host of neurological problems such as convulsions, muscular rigidity, tremors and inability to walk or talk, and they are usually placed in institutional care. In North America and Europe phenylketonurics make up nearly 1 per cent of mentally defective populations. Fortunately there is a preventive course of action. Phenylketonuria can be detected at birth by a routine biochemical test. If a baby is so diagnosed, then the immediate and long-term administration of a phenylalanine-low diet will markedly reduce body levels of phenylalanine and thereby halt the brain damage. Phenylketonuric children maintained from the earliest age on a low-phenylalanine diet should develop normally.

The prevalence of phenylketonuria is about 1 in 12 000 although there is a wide variation among different population groups. Its highest frequency is among North European Caucasoids; it is uncommon among Mongoloids, and exceptionally rare in African Negroids and Ashkenazi Jews.

Hypopigmentation

Phenylketonuria was first described by the Norwegian, Følling, in 1934 and he made the astute observation that almost all the affected patients had blond hair, blue eyes and fair skin. This finding has been amply confirmed in subsequent surveys, and indeed dilution of pigment is regarded as a characteristic feature of the disorder. While the great majority of phenylketonurics are blue-eyed blonds, the coloration of a particular person has to be judged against that of unaffected members of the family and ethnic group. Japanese phenylketonurics have dark brown hair which contrasts with the normal black hair of the population, but blond hair and blue eyes have been noted in phenylketonuric children from dark-skinned Mediterranean families. The lighter hair colour of these patients compared with their unaffected siblings has been demonstrated objectively by reflectance studies (Cowie & Penrose, 1951; Roberts, 1977). The hair of phenylketonurics darkens with age, but at a slower rate than that of controls (Cowie & Penrose, 1951). The lowering of blood phenylalanine concentrations by diet causes an accelerated darkening of hair colour. There is suggestive (but inadequate) evidence that, unlike oculocutaneous albinos, phenylketonurics have deficient neuromelanin in the pigmented brain areas (substantia nigra and locus caeruleus) (Corsellis, 1953; Fellman, 1958; Crome, 1962).

There have been reports of two phenylketonuric patients who showed neither hypopigmentation nor low intelligence. These have given rise to the belief that the lighter the complexion in phenylketonuria the lower the IQ (Ortonne, Mosher & Fitzpatrick, 1983), and to the suggestion that abnormal phenylalanine levels cause parallel biochemical disturbances within the melanocyte and the brain neurone. However, this skin colour–IQ association has not been substantiated and the issue remains unsettled.

The mechanism for the pigmentary dilution in phenylketonuria has not been finally resolved. The favoured explanation is that the raised phenylalanine levels in the body inhibit the formation of melanin. It has been demonstrated that the concentrations of phenylalanine found in untreated phenylketonuria inhibit mammalian tyrosinase (Miyamoto & Fitzpatrick, 1957a). The authors of a more recent study (Farishian & Whittaker, 1980) have postulated that the phenylalanine-related reduction of melanin biosynthesis is due not to inhibition of tyrosinase activity but to a decrease in tyrosine uptake by melanocytes.

Although phenylketonuria has dramatic effects on pigmentation it is not the only metabolic disorder to do so. There is an abnormality of methionine metabolism known as *homocystinuria*. This is most commonly due to a deficiency of an enzyme, cystathionine synthetase, and the majority of affected patients have blond hair, blue eyes and fair skin (Ortonne *et al.*, 1983), even those of Mediterranean extraction. In some of these patients the hair has darkened with the administration of vitamin B_6 (pyridoxine). In another inherited metabolic disorder, *cystinosis,* there is an accumulation of the amino acid cystine within the skin and eye, with resultant hypopigmentation of these tissues giving an appearance reminiscent of oculocutaneous albinism, but without the iris translucency (Stenson, Siegel & Carr, 1983).

Vitiligo

Vitiligo is an acquired, circumscribed and progressive loss of pigment, most commonly affecting skin and hair but occasionally involving other melanocyte-bearing areas, such as the uveal tract of the eye and the retina. The disease may be localized or widely spread, and it manifests with discrete milk-white patches on the skin (varying greatly in size and shape) that may either be scattered or fused into one another to form large confluent areas. The dappled appearance of the patient is bizarre,

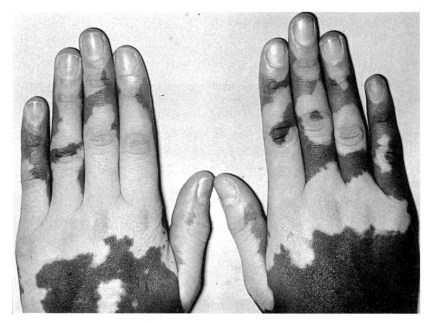

Fig. 9.7 Vitiligo affecting hands of South African Negroid patient. Note fairly symmetrical distribution of lesions. (Courtesy of Professor Norma Saxe.)

and cosmetically vitiligo is among the most disfiguring disorders, especially in darkly pigmented patients in whom the white patches stand out starkly against normal skin (Fig. 9.7). The former Prime Minister of India, Nehru, regarded vitiligo as one of the conditions requiring the most urgent attention in his country.

Vitiligo has a very long history, and descriptions of it occur in the Indian sacred book, *Atharva Veda*, which dates back about 3000 years. References to 'white skin' blemishes are found in the historical writings of many religions, including Buddhism, Islam and Shintoism (Ortonne *et al.*, 1983). In the Old Testament (Leviticus, Chapter 13) the Hebrew word for certain skin disorders (including depigmentation) became translated in the Greek version of the Hebrew Scriptures (250 BC) to mean leprosy. This was entirely incorrect as the skin diseases described in the Bible were neither contagious nor in any way similar to the current concept of leprosy, but considerable damage resulted because patients with vitiligo were severely stigmatized as lepers and outcasts – and still are today in several countries. This prejudice has caused vitiligo patients acute self-consciousness and subjected them to ostracism.

Vitiligo is a disease affecting all ethnic groups (its average prevalence being about 1 per cent with a range of 0.1–8.8 per cent (Ortonne *et al.*, 1983)). Statistics reveal that it is commoner in the Indian subcontinent than in Europe, but this may only reflect the desperate need of Indian patients for treatment because the condition brings such devastating social consequences. There is often a family history of vitiligo (one-third of patients have one or more affected relatives) and this suggests that a dominant gene may be involved. The disease manifests at any stage from birth to old age, but the onset is usually between the ages of 10 and 30 years. It tends to occur more frequently in dark-haired, brown-eyed subjects and in those who tan well, but again this may only indicate that dark-complexioned patients are more disturbed by the onset of vitiligo and come forward to be counted and treated. Once the lesions appear, they tend to spread (sometimes rapidly) and there is usually little chance of effective repigmentation. Vitiligo can occur anywhere on the skin but it often declares itself (ironically) in normally hyperpigmented areas (e.g. face and limbs) and at sites of friction and trauma.

Causation

No single factor is known to induce vitiligo but many patients have observed that specific events preceded the vitiligo (e.g. physical injury, sunburn, pregnancy, emotion). The contribution of psychological factors has often been noted and there are anecdotal reports of the condition developing overnight after an acute emotional trauma, although it is always problematic with vitiligo to gauge the degree of psychological causation. There are additional considerations, such as the patient's emotional stability before the disorder appeared or the psychological effects of the vitiligo itself (probably the most usual way in which the disorder and emotion are linked). A further difficulty is that the coincidence of a stressful event with the onset of vitiligo does not spell causality. There are many cases that arise without any history of stress or trauma. The characteristic finding in an area of vitiligo is the total absence of melanocytes, melanin and tyrosinase activity; otherwise the epidermis and dermis are normal.

The cause of vitiligo is unknown but there are several hypotheses. The most popular is that the condition is due to an immunological process (auto-immunity) whereby the body generates its own antibodies (auto-antibodies) to melanocytes which are then destroyed in an antigen–antibody reaction. Various other diseases – hyperthyroidism, Addison's disease, pernicious anaemia and diabetes mellitus – are associated with autoantibodies, and it is interesting that there is an abnormally high incidence of vitiligo in patients with these diseases. In Addison's disease, for example, patches of spreading vitiligo co-exist with the areas of hyperpigmentation. Another intriguing lead is that patients with malignant melanomas who develop vitiligo seem to be more resistant to the life-threatening effect of the malignancy; they survive longer than their counterparts without vitiligo. It is paradoxical that two biologically opposite activities, namely, the uncontrolled proliferation of melanocytes and the destruction of melanocytes, should occur in the same patient. Perhaps the immune response partially obliterates the melanoma cells but does so at the expense of innocent melanocytes.

Anti-melanocyte antibodies have been detected in vitiligo patients but it is not known what their role is in the pathogenesis of the condition or, indeed, whether they have any significance at all in the disease process (Ortonne *et al.*, 1983; Bystryn & Pfeffer, 1988).

One type of vitiligo shows a segmented distribution, i.e. it involves areas of the skin supplied by the cutaneous nerves. In this type nervous factors are possibly the most important, and one proposal is that neurotoxic substances liberated from nerve endings inhibit melanin biosynthesis and may even destroy the melanocyte.

A third explanation is that melanocytes actually destroy themselves ('self-destruct hypothesis') when they lose their natural protective mechanism against the melanin precursors (phenol or catechol derivatives) which are selectively cytotoxic to melanocytes (Lerner, 1971). This idea was based on the clinical finding that vitiligo often occurs in areas of hyperpigmentation.

Treatment

Patients with vitiligo must protect depigmented areas against sunburn with the appropriate sunscreen agents. There are also cosmetic preparations which can dye or camouflage the affected areas. In patients with severe vitiligo and relatively little normal skin, the approach has been to apply chemical depigmenting agents to the intervening normal areas so as to achieve a uniform coloration. A more specific therapy is the use of the photosensitizing psoralen agents in combination with long-wave ultraviolet light (UV-A) (see p. 53). This treatment (known by the acronym PUVA) probably acts by causing proliferation of epidermal melanocytes with enhancement of tyrosinase activity and increased melanogenesis.[2] Gradually there is repigmentation of the affected areas, but PUVA takes at least 1–2 years of continuous therapy to be effective and it thus requires an enormous commitment of time and stamina. PUVA may have unpleasant effects (e.g. gastrointestinal symptoms, darkening of the normal skin) and the success rate is of the order of 50–70 per cent. These factors demand careful patient selection, and perhaps the most important criterion in any particular patient is the strength of his or her motivation. Although PUVA therapy is relatively recent, the ancient Indian text, *Atharva Veda*, mentions a plant which is able to restore normal skin colour. This plant contains psoralen as its active principle!

Recently, the administration of oral phenylalanine together with UV-A or sunlight irradiation was found to induce repigmentation in about 80 per cent of patients with vitiligo (Kuiters *et al.*, 1986). A more exciting development is the possibility of transplanting autologous melanocyte cultures (i.e. those derived from the normal pigmented skin of vitiligo patients) to the affected areas which are devoid of melanocytes (Lerner *et al.*, 1987).

Piebaldism
This is an uncommon condition (inherited as an autosomal dominant) which may closely resemble vitiligo. It is sometimes erroneously known as 'partial albinism', and it is characterized by the presence of a congenital white forelock which is triangular in shape and situated symmetrically in the midline of the forehead. In addition, there are areas of depigmented skin over the body which are identical to vitiligo microscopically in that there is an absence of melanocytes, melanin and tyrosinase activity. However, the difference is that in piebaldism these areas incorporate within their boundaries islands of normal or hyperpigmented skin. Moreover, while vitiligo is an acquired and progressive disorder the lesions in piebaldism are present at birth and remain static throughout life. The latter characteristic makes piebaldism eminently suitable for autologous melanocyte transplants.

Other genetic hypopigmentation disorders
Waardenburg syndrome
This is a rare disorder, determined by dominant inheritance, which consists of a number of features among which are a white forelock, a broad root of the nose, heterochromia of the irises (one eye differing in colour from the other), congenital deafness and depigmented skin patches (with absent melanocytes). It is not necessary that all of these signs be present for the diagnosis of the syndrome.

The most disabling component of the Waardenburg complex is the deafness. During a diagnostic survey of deaf children attending special schools in Southern Africa, 3 per cent had Waardenburg syndrome (Sellars & Beighton, 1983). These children included Negroids, Caucasoids and those of mixed ancestry, and a surprising finding was that some profoundly deaf Negroid children has irises of a very unusual, striking blue colour (Fig. 9.8). These children lacked any other stigmata of the condition, although several had first-degree relatives with the Waardenburg syndrome.

The occurrence of blue eyes in Negroids may be part of the same syndrome. The blue-eyed Negroid has always been a curiosity because it was believed that dark skin and hair pigmentation are invariably accompanied by dark pigmentation of the eyes. Gates (1938), however, described several members of a family of pure Singhalese descent in Ceylon

Fig. 9.8 South African Negroid boy with Waardenburg syndrome is deaf and has eyes of a remarkably clear blue colour, in sharp contrast with normal dark brown-black eye colour of South African Negroids. (Courtesy of Professor Peter Beighton.)

(now Sri Lanka) who had blue eyes. Tsafrir (1974) has presented a series of South African Negroid family pedigrees in which two or more members (of pure Negroid descent) had two normal but completely blue eyes. There was no evidence of the Waardenburg syndrome or any other pathology. She concluded that the blue eyes probably arose as an independent mutation which was thereafter transmitted by an irregular dominant mechanism. The latter is unusual because in Caucasoids light eyes are generally controlled by recessive genes.

The Waardenburg syndrome is an example of one of several rare genetic conditions embracing both deafness and pigmentary abnormalities. Melanocytes, which are derived from the neural crest, populate not only the epidermis but several other sites including the inner ear (see pp. 76–7). The signs of a condition like the Waardenburg syndrome are compatible with a defect in the neural crest, the derivatives of which fail to differentiate or to migrate to their target organs (Schrott & Spoendlin, 1987).

Tuberous sclerosis

This is an uncommon condition which characteristically produces a triad of signs: mental retardation, convulsions and multiple small nodules on the face. The interesting aspect of tuberous sclerosis is that the advent of this triad is preceded by the appearance of leaf-shaped ('ash leaf') hypopigmented spots (Fig. 9.9). This enables the diagnosis of tuberous sclerosis to be made at birth (before any of the other features emerge) if these specific lesions can be detected. The white spots occur in the great majority of affected babies, their average number being about five per patient (Ortonne *et al.*, 1983). The ability to establish a diagnosis of tuberous sclerosis at birth is important because it will allow for the planning of the child's future needs. In addition, some cases of tuberous sclerosis are genetically transmitted and an early diagnosis can greatly facilitate genetic counselling.

Electron microscopy of the spots reveals an unusual picture with a normal number of melanocytes but a reduction in the number, size and melanization of melanosomes (Jimbow *et al.*, 1975). Most of these melanosomes are aggregated in complexes in the keratinocytes, and this also applies to Negroid patients with the condition. (Melanosomes in Negroid subjects (see p. 17) are typically distributed in the single state in keratinocytes.) The latter observation is a further confirmation that the pattern of melanosome distribution is not determined by race but by the size of the melanosomes.

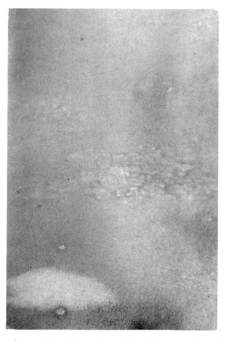

Fig. 9.9 Skin of lower back in patient with tuberous sclerosis, showing leaf-shaped ('ash leaf') hypopigmented macule in left lower corner. Leathery area in centre is shagreen patch, also a common feature of the disorder. (Courtesy of Professor Norma Saxe.)

Chemical hypopigmentation

A large number of chemical compounds can induce depigmentation and the resultant effect resembles vitiligo. The best-known of these agents are *hydroquinone* and the *monobenzyl ether of hydroquinone*.

Monobenzyl ether of hydroquinone was used in the rubber industry to prevent the oxidation and spoiling of rubber during its processing. Its action on the skin was first observed in factory workers who had regularly worn rubber gloves containing the compound. They developed depigmentation (especially striking in Negroids) on the hands and half-way up the forearm, the area corresponding to glove contact. Depigmentation of the penis has occasionally followed the wearing of rubber condoms. In many instances the depigmentation due to monobenzyl ether of hydroquinone was permanent, probably because the compound actually caused destruction of melanocytes. Hydroquinone is less potent than its monobenzyl ether and, while it has the ability to destroy melanocytes in experimental animals, it is unlikely to produce irreversible changes (Ortonne *et al.*, 1983).

Bleaching creams

Monobenzyl ether of hydroquinone is used in the treatment of extensive vitiligo to bring about the irreversible depigmentation of the intervening patches of normal skin. Because of its toxicity the ether is no longer used in the treatment of other pigmentary disorders, for which hydroquinone is the preferred agent. Hydroquinone produces satisfactory bleaching of hyperpigmented skin (e.g. chloasma) with fewer adverse effects than the monobenzyl ether.

In South Africa and other African countries there has been an epidemic in the use of skin-lightening cosmetics. The craze for suntanning and sunscreen preparations among fair-complexioned Caucasoids is matched by the passion among many Negroids (particularly women) for preparations that will lighten and brighten their skins and thereby (they believe) enhance their attractiveness. This attitude has been exploited by cosmetic marketeers whose subtle advertising reinforces the concept that light-skinned people are more successful. The result has been the promotion of a multi-million dollar industry in over-the-counter bleaching creams, an industry that has (not surprisingly) ignored the pleas of the medical profession.

Unfortunately, these bleaching creams are not safe. The earlier preparations were withdrawn from distribution because they contained mercury salts which were absorbed through the skin and caused kidney damage. Then monobenzyl ether of hydroquinone was incorporated into bleaching creams, with a resultant outbreak of *leucomelanoderma* (Dogliotti *et al.*, 1974). This condition, which was seven times more frequent in females than in males, was a cosmetic disaster for a dark-skinned individual – it manifested as an ugly pattern of patchy depigmentation mingled with mottled hyperpigmentation. However, even when hydroquinone was substituted for the monobenzyl ether the situation hardly improved. The habitual application of hydroquinone-containing bleaching creams also generated a new skin disease of the face (termed *ochronosis*) (Findlay, Morrison & Simson, 1975; Phillips, Isaacson & Carman, 1986). After a year or more of effective use, these preparations lead to a darkening and coarsening of the skin over the cheekbones followed by the appearance of fixed, pitch-black papules (likened to lumps of caviar) (Fig. 9.10). Among black South Africans, cases of ochronosis constitute a significant proportion (about one-third) of patients attending dermatological clinics. There is no satisfactory treatment for the disease: the lesions are due to progressive, degenerative changes in collagen. There is no deficiency of melanocytes, and the skin is probably hyperpigmented due to increased melanin in the dermis (Phillips *et al.*, 1986).[3] Initially, it was considered that hydroquinone caused skin problems only if present in high amounts and that, at concentrations of 2 per cent or less, hydroquinone in bleaching creams would be safe. However, even with these lower concentrations, ochronosis frequently develops and the only legitimate public health policy – certainly in

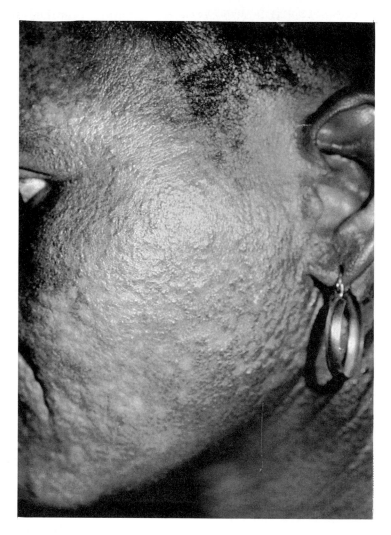

Fig. 9.10 Ochronosis as a result of prolonged application to face of hydroquinone-containing bleaching creams. Note coarsening of skin over cheekbones and appearance of caviar-like papules. (Courtesy of Professor Norma Saxe.)

countries where skin lighteners are abused – is to ban them from the over-the-counter market, a step taken by South Africa in 1990.

Sudden whitening of the hair

The alleged phenomenon of sudden whitening of the hair is a topic that has long provoked controversy and conflict in the spheres of legend, history and science. It is the example *par excellence* of the startling effect of psychological trauma on the pigmentary system.

It is important to distinguish between sudden whitening of the hair and greying of the hair. The latter is the progressive change of hair colour that accompanies aging (see p. 22). Greying of the hair may become notice-able in the twenties although the usual age of onset in Caucasoids is about 35 years and in Negroids a decade later. The process is a slowly progressive one, involving the replacement of pigmented by hypopig-mented hairs. 'Sudden whitening' denotes the abrupt (virtually over-night) appearance of totally depigmented hair and, according to numer-ous historical sources, it is usually precipitated by the occurrence of catastrophic emotional stress. The English language has taken up this cue in the everyday expression, 'to give one grey hairs'. And, as Oscar Wilde mischievously noted: 'Her capacity for family affection is extraordinary. When her third husband died, her hair turned quite gold from grief'!

Marie Antoinette's hair is reputed to have turned white after the insults and abuse suffered by her and the Royal family during their return to Paris after an unsuccessful attempt to flee the country and the French Revolution. The hair of Ludovico Sforza, the fifteenth-century ruler of Milan, rapidly turned white after his capture by Louis XII; the hair and beard of Sir Thomas More became white on the night prior to his execution. There are numerous other examples, not only from history (Jelinek, 1972) but also from the medical literature. In a review of the latter (Ephraim, 1959), there were 26 cases of rapid or sudden hair whitening of which some were preceded by excitement, fright or mental stress (e.g. terrifying battle experiences, shipwrecks, railway catastrophes) and others by mental or neurological disorders (e.g. seizures, strokes, brain injury).

There have been many theories as to the mechanism of sudden hair whitening. Some scientists have outrightly rejected such occurrences and relegated them to mythology despite the reports of reliable observers. The problem which has baffled the medical profession is how pre-existing and stable melanin pigment can vanish over such a short period of time. Greying is a gradual development which occurs in the period necessary for the physiological growth of individual hairs. Metchnikoff (1901), the

Nobel Prizewinner in Physiology and Medicine, communicated the idea to the Royal Society of Medicine that in sudden whitening hair pigment was absorbed by phagocytes, transported away from the hair bulb and deposited in the connective tissue. Another popular theory claimed that, at the moment of psychic trauma, air bubbles entered the hair shaft in the presence of pigment and produced a white appearance through the reflection of light (Jelinek, 1972). Neither of these proposals had the slightest basis in fact!

The current view is that, although very rare, sudden hair whitening can and does occur – despite the incredulity of scientists. There are too many well-documented authentic instances for it to be dismissed as folklore. One report (Helm & Milgrom, 1970) described a patient whose hair suddenly appeared to have turned white although, as the disease progressed, he exhibited patches of complete hair loss, a condition known as *alopecia areata*. Initially, he had predominantly brown hair with less than a 5 per cent admixture of white hairs, but over 3 days he lost almost all the brown hair so that his remaining scalp hair looked white.

Thus, the selective loss of *pigmented* hair seems to furnish a reasonable explanation for the enigmatic phenomenon of sudden whitening. For unknown reasons, the hair loss in alopecia areata frequently confines itself to the pigmented hairs with the white hairs remaining intact (Jelinek, 1972). Most middle-aged people have a mixture of pigmented and white hairs and, if the former are preferentially shed, the resultant hair covering, albeit sparser, would be white. Alopecia areata may occur explosively – in a matter of days and to an alarming degree – and it is sometimes associated with acute emotional trauma, although the role of psychological stress in directly causing this disorder is uncertain.

Notes

1 The Reverend W. A. Spooner (1844–1930), Warden of New College, Oxford, was an albino (probably tyr-neg) and, according to his biographer (William Hayter (1977). *Spooner: A Biography*. London: W. H. Allen), he looked like 'a white-haired baby' throughout his life. His short-sightedness was severe and, before writing his final examinations at Oxford, he had to apply for a special window seat.

2 The administration of melanocyte-stimulating hormone to patients failed to repigment the vitiliginous areas although it produced considerable darkening of the normal skin.

3 It appears that the melanocyte itself is essential in the development of ochronosis. A case is reported of a Negroid woman with vitiligo patches on the face who used a hydroquinone-containing bleaching cream to lighten her normal skin. She subsequently developed ochronosis but the areas of vitiligo were spared (Hull, P. R. & Proctor, P. R. (1990). *Journal of the American Academy of Dermatology, 22*, 529–31).

10   *Skin colour and society: the social–biological interface*

W. E. B. DuBois once remarked that the problem of the twentieth century was the problem of the colour line; and even in the closing years of this century there is little doubt that the impact of skin colour continues to be profound, whether at the macro-level of global politics or the micro-level of a person-to-person transaction. Like three other exceedingly small entities – the atom, the ovum and the AIDS virus – the melanosome still has a place on the agenda of human catastrophe. It would therefore be a glaring omission in a book such as this to overlook the immense interaction between skin pigmentation and the psychosocial dimensions of human behaviour.

Legends, symbolism and culture

The colour of the skin, hair and eyes has intrigued people from time immemorial, as it has also engendered curiosity about the reasons for colour differences between human populations. In prescientific eras much of the thinking on the subject was based on mythology or primitive religious concepts. The well-known scriptural interpretation from Genesis blamed blackness on a curse delivered by Noah to his son Ham as a punishment for having gazed on him when he lay naked and drunk in his tent. The Ancient Greeks narrated that Phaeton, the son of Helios (god of the Sun), successfully coaxed his father to allow him to drive the fiery chariot of the Sun for one day. His maladroitness caused him to lose control of the reins so that the chariot came too close to the earth in one region (Ethiopia), burning the people there black, and was too far from the earth in other regions, turning the inhabitants there pale from cold.

In the English language whiteness has been associated with beauty, purity, goodness, innocence and brightness. Blackness has symbolized death, deceit, filth, disgrace and a host of other adverse descriptions, the essence of which has been incorporated into the language in such words and phrases as blackguard, blackball, black flag, black list, black market and black sheep. Mason (1968) has pointed out that these connotations of black and white for evil and goodness respectively apply also to other

166

languages (e.g. Latin, Greek and Sanskrit). However, associations with the word 'white' are not always favourable, as in white feathers, whited sepulchre, white-livered, whitewash and white flag.

This colour symbolism of white–good/black–bad is prevalent throughout the world and pervades peoples as diverse as the Ndembu of Angola (Turner, 1966), the Dogon and Yoruba of West Africa, the Malagasy of Madagascar, the Sakai of the Malay Peninsula, the Cherokee of North America, the Chuckees of Siberia (Gergen, 1968), and also the French, Italians and Germans (Best, Naylor & Williams, 1975; Best, Field & Williams, 1976). It is important to emphasize, however, that there are exceptions: to the Shona of Zimbabwe and the Sandawe of Tanzania black represents the rain-bearing clouds which usher in the wet season, and it therefore becomes an auspicious colour, especially in drought-prone regions (Turner, 1966). In Egypt black was the colour of regeneration (probably relating to the black colour of fertile soil), and Egyptian statues were coated in black resin to enhance their potency. But, apart from these few situations, the positive and negative attributions of whiteness and blackness respectively appear to transcend ethnic, religious and cultural barriers; and in a sense they are 'primordial'. This near universality of the white–good/black–bad connotation has been explained not on any racial basis but on the diurnal nature of *Homo sapiens* (Best *et al.*, 1975, 1976). It has been proposed that young children develop their initial liking for white over black as a result of their early experiences with light and darkness. The latter brings with it isolation and fear whereas lightness is the bearer of nourishment, activity, stimulation and human contact. This preference for light over darkness then attaches to white and black, and thereafter the phenomenon becomes incorporated in, and modified by, a variety of socio-cultural and religious influences. Some support for the hypothesis derives from the finding that the degree of white–positive/black–negative bias is significantly correlated with children's reported aversions to the dark of night and to thunderstorms (Boswell & Williams, 1975).

The colour associations of black and white have been readily assimilated into aesthetic judgments. A fair or white skin was equated with beauty even in Biblical times. The English word 'fair' (from Anglo-Saxon, *faeger*) assumed the dual meaning of beautiful and light-skinned, and indeed fairness of complexion became synonymous with feminine beauty from about the thirteenth century. It is also significant that the entire history of Western painting shows a deliberate whitening that transformed Christ – his dark hair and eyes in the early pictures gradually giving rise to a blond, blue-eyed appearance, so that the representation of the incarnation of God was as remote as possible from darkness or blackness (Bastide, 1968). Conversely, the devil is portrayed in early Christian art as a black man in black clothing, and thus blackness was linked to satanic practices (black magic).[1]

The concept of fairness and beauty is not restricted to English or European norms. In India a lightish skin colour is a desirable component of beauty, and Indian languages also use the words 'fair' and 'beautiful' synonymously. The ideal bride has a light complexion, a feature that is noted in the matrimonial columns of Indian newspapers. Fair-skinned children (particularly daughters) are highly valued within Indian families. A similar scenario prevailed in the West Indies where colour prejudice was directed much more against dark or black women than against similarly coloured men (Lowenthal, 1972). Thus, the successful black man sought a light-skinned wife 'to raise the colour of the family'. Whiteness of the skin is a feature of beauty in Japan and it is linked to purity and chastity (Wagatsuma, 1968). The skin of east Asian peoples tends to tan readily – so that Japanese women, for example, can only maintain their lightness by the total avoidance of sunlight or by the constant use of parasols together with the application of special 'bleaching' powders to the face, neck and upper chest. Japanese districts with a great deal of cloud cover and little sunlight are renowned for their fair-skinned beauties whereas, as the proverb puts it, 'In the provinces where one can see Mount Fuji one can hardly see beautiful women.' In Africa similar perceptions are encountered. To the Ibos of Eastern Nigeria yellow-skinned girls and women are highly regarded and command higher bride prices (Ardener, 1954). Women with light skins are admired among the Lunda, Luvale and Chokwe of Zambia (White, 1954) – and very black skins are believed to indicate inherited witchcraft. Colour consciousness in marriage is also a feature of certain groups of Solomon Islanders (Baldwin & Damon, 1973).

van den Berghe & Frost (1986) identified 51 human societies throughout the world (86 per cent outside Europe and North America) where skin colour constituted an indigenous criterion of beauty or attractiveness. In 47 of these the sexual preference was for a lighter skin colour but this applied only to women in 30 of the societies. The influence of colonialism or Western aesthetic values was discounted as a major determinant of the preference. Because almost all post-pubertal females, irrespective of ethnic group, are lighter in pigmentation than their male counterparts (see p. 112), van den Berghe & Frost proposed that this particular sexual dimorphism developed to facilitate the attraction of males by females (sexual selection).

The advantages of possessing a light skin are so great that the sale of skin-lightening creams has become an enormous commercial enterprise, despite the undisputed hazards of such preparations (see p. 161).[2] Most of these compounds now employ hydroquinone as the active ingredient, but at one time they contained mercury which had the potential to cause kidney disease (nephrotic syndrome). Barr *et al.* (1972) reported on a series of Kenyan patients with mercury-induced nephrotic syndrome who differed from other medical inpatients at the same teaching hospital: they were younger, the great majority were female and a markedly higher proportion (three to four times greater) spoke English. This profile indicated that young, sophisticated Kenyan women had the need to lighten their skin colour artificially in order to achieve or maintain a high level of social acceptance and sexual desirability.

There is a high prevalence of albinism (1 in 3900) in South African Negroids (Kromberg & Jenkins, 1982), a surprising finding in an environment where the albino would be at a considerable disadvantage from high UV (see p. 144). The parents of South African Negroid albinos (who are not themselves albino but are heterozygous carriers of the albinism gene) have a lighter skin colour than controls (Roberts, Kromberg & Jenkins, 1986) (see Figs. 9.5 and 9.6). This lighter coloration of the heterozygote might bring social advantage to that person and with it preference as a marriage partner. The operation of both sexual selection and assortative mating (see p. 113) would facilitate transmission of the gene for albinism and thereby promote the high occurrence of the disorder.[3]

Fairness of skin is not only an asset with regard to marriageability but it is also esteemed as an attribute within the wider society. In India, light-skinned people are very common among established landowning and aristocratic families, and indeed the Indian caste system generated a variety of stereotypes – the upper castes being represented as fair and the lowest castes as dark. Classical Hindu texts in fact assigned colour to the four orders of humankind in a hierarchy from light to dark – white to the Brahmans (priests) at the top, yellow and bronze to the middle groups of Kshatriyas (warriors and rulers) and Vaisyas (merchants) respectively, and black to the Sudras (artisans) at the bottom. The Sanskrit word 'varna', which is used to denote these orders, also means 'colour'. (The 'untouchables' (Harijans) fell outside the varnas.)

The skin colour of six endogamous groups in north India was measured by reflectance spectrophotometry (Jaswal, 1979). This study confirmed that the Brahmans had the lightest pigmentation (highest reflectances) at an unexposed area and that generally, in the same geographical region, there was some correspondence between skin colour and social status (caste hierarchy) (see Table 7.3, note b). Similarly, among Japan students, the upper class group had higher mean reflectances than did the middle and lower groups (Hulse, 1967), and this supported the idea that social selection had favoured the concentration of genes for lighter skin colour in the upper classes (see Table 7.3, note e). The effect of skin colour on social class was investigated by reflectance spectrophotometry in Anglo-Americans and Mexican-Americans living in San Antonio, Texas (Relethford *et al.*, 1983). While social class had no influence on skin colour variation for Anglo-Americans there was a highly significant effect for the Mexican-Americans, in that the greater the native American ancestry – and the darker the skin colour – the lower the socio-economic status.[4]

In the Caribbean and Latin America skin colour *per se* has been less important in determining social standing than power, wealth and prestige. The Brazilian saying, 'money bleaches', emphasizes the point that individuals of financial means could override the status determined by their skin colour. Prosperous or successful people were thought of 'as if' they were white, irrespective of their actual appearance, and lower class people were seen as black.

The colour problem: historical and philosophical aspects

Unfortunately, the myths and legends about skin colour and the socio-cultural perceptions of it have been far from innocuous, and in certain societies they developed into a firm and fixed set of beliefs which held white people to be innately superior to black people. The intensity and passion which have fuelled these convictions have been translated into effective action programmes in which blacks have been abused, mistreated and oppressed.

Nowhere were the above attitudes more harshly fashioned than within the context of North American slavery, the details of which have been meticulously chronicled by Jordan (1968). His book gives appalling insights into the atrocious practices inflicted by white Americans on their black slaves. It is of great interest that, although slavery was practised by the Spanish and Portuguese in the New World, it was very much less oppressive and racist than that of North America. The probable reason is that the indigenous populations of Spain and Portugal had been under the aegis of the Moors until the fifteenth century. The Moors were darker-skinned, more cultured and better educated than the subject peoples of the Iberian peninsula. Thus, when Henry the Navigator imported black slaves into Portugal in substantial numbers they were integrated into the local community with little, if any, colour prejudice.

It is also pertinent to reflect on whether the American antipathy to blacks had parallels in the ancient Graeco-Roman world. This question had been reviewed by Snowden (1970) who concluded that, apart from some colour-orientated superstitions, blacks (Ethiopians) were favourably received by the ancient Greeks and Romans. Their piety, justice and wisdom were respected; their physical characteristics were regarded as a response to climate and geography, and there were neither pejorative attributes related to their blackness nor dogmas of white superiority. Indeed, an absence of colour prejudice and biological racism prevailed and, if individuals were slaves, then they were treated well or badly regardless of ethnicity.

Lewis (1970) has examined the Muslim attitudes to blackness and blacks, and found that they embraced those adverse racial stereotypes that would later be taken up by Europeans and antebellum Americans. Blackness had the insinuation of ugliness and inferiority, and this expressed itself in social discrimination against dark-skinned people. In the Muslim world there were white and black slaves but the whites (Jews and Christians) were rarely used for hard labour and were much more often favoured and promoted, even to the rank of general or provincial governor. Muslims viewed whiteness as a mark of superiority, possibly in part because of Noah's curse on Ham, which Muslim legend also interpreted as a damnation to blackness and to eternal slavery. It should be noted, however, that, unlike Judaeo-Christian and Hindu texts, the Koran itself is remarkably free of racial or colour prejudice. The Koran recognizes differences between peoples but these are not translated into racial terms.

From the historical perspective, a major impetus towards racist thinking in the eighteenth and nineteenth centuries derived from contemporary biological science. Linnaeus, in 1758, had set the scene by his skin colour classification of humankind which he coupled with personality typologies. Thus *Homo Europaeus* (white) was 'lively, light, inventive and ruled by customs'; *Homo Americanus* (red) was 'tenacious, contented, choleric and ruled by habit'; *Homo Asiaticus* (yellow) was 'stern, haughty, avaricious, ruled by opinions'; and *Homo Afer* (black) was 'cunning, slow, phlegmatic, careless and ruled by caprice'.

There was another concept which also had a profound impact on eighteenth-century biology and that was the 'great chain of being'. This concept envisaged that all living organisms formed continuous gradations, like the rungs of a great ladder, which spanned the spectrum from the amoeba to the angels. By 1850 two ideas had become firmly held: one was that the races themselves slotted into this great chain of being, the Caucasoid being on top and the Negroid at the bottom; the other was the shift in paradigm from monogenism to polygenism (Stepan, 1982). Monogenism conceived that the various races originated from a single source and were therefore interrelated (an 'Adam and Eve' model); polygenism postulated that human races were separate biological species. The merging of the chain of being and polygenistic theories led directly to the concept that the emergence of Negroids in Africa was closely linked to the origin of the great apes.

James Hunt, founder and president of the Anthropological Society of London, became a leading British exponent of scientific racism in the 1860s (Lorimer, 1978). In 1863 he published an essay in which he concluded that the Negroid was a distinct species from the Caucasoid, intellectually inferior and more akin to the ape. Although statements such as those of Hunt and others were strongly rejected by serious intellectuals of the day they did make an impact on mid-Victorian society, especially as they carried the stamp of learned authority. In the earlier part of the nineteenth century there had been no clear-cut pattern of racial discrimination in England, although the experience of British slavery (particularly in the previous century) had seriously damaged the self-concepts of blacks and had created negative attitudes towards them. In this period blacks were treated in terms of their social standing rather than their race, and the English were prepared to accept gentlemen as gentlemen. But as the century advanced the idea of the inherent superiority of a white skin asserted itself and blacks, irrespective of their education or position, were lumped into one category and excluded from the ranks of gentlemen (Lorimer, 1978). This change in English public opinion was catalysed by the Jamaica Insurrection of 1865 and by the growing climate of scientific racism. Indeed, it is remarkable how vehemently the doctrine of Negroid inferiority was accepted by sectors of the scientific community up to and into the twentieth century. The subject has been carefully researched by Gould (1981) and Stepan (1982).

The most ominous incentive for the subsequent abuse of skin colour as a criterion of human worth stemmed from the writings of Arthur de Gobineau (1816–1882) which had a profound effect on the intellectual ferment of the late nineteenth century. Gobineau divided humankind into white, yellow and black; all that was great and glorious in the world was credited to the creative talents of the white race – and especially to its mythical Aryan branch, which was held to have been responsible for all the principal civilizations in the world's history.

Gobineau thus espoused the doctrine of 'Aryanism'. His Aryan race was a 'superior' caste; the pure-bred, select and privileged minority, born to govern and direct the destinies of the 'inferior' cross-bred masses in any nation. (The word *Aryan* (meaning 'noble' or 'pure') derives from the Sanskrit legends in which the light-skinned conquerors, the 'Aryas', who emanated from the north 3500 years ago, dominated the indigenous dark-skinned 'Dasyas' of the Indian peninsula.) Subsequently, the followers of Gobineau ascribed a cluster of characteristics to the exclusively Aryan type – tallness, blue eyes, fair hair and long heads, with the associated psychological features of virility, innate nobility, responsibility, tenacity of will and leadership. Houston Stewart Chamberlain (1855–1927) elaborated on Gobineau's ideas and alleged that the Aryan race (and especially its Teutonic division) constituted one which was predestined to become the political and cultural leader of the world for centuries ahead. This ideology led straight to the 'master race' doctrine of Adolph Hitler who, in *Mein Kampf,* proclaimed that when the Aryan had mixed his blood with those of inferior peoples then this caused the ruin of the civilizing races. Hitler's concept of racism became the inspiration of the National Socialist Party and the spiritual basis of the Third Reich. German blood had to be purified by eliminating alien minorities and by exterminating the Jews.[5] This was the grandiose racial myth that built the concentration camps and culminated in the genocide of the Holocaust.

The colour problem in the twentieth century

The legacies of the past have reasserted themselves in the twentieth century, and even up to the present time colour prejudice and racial discrimination have continued to plague societies throughout the world. In the United States of America blacks suffered grievously from oppression and persecution – lynchings, the Klu-Klux-Klan and segregated public facilities (Jim Crowism) were all part of the tapestry of American racism. One of the most hypocritical examples of institutional racism was the segregation of the United States armed forces up until 1948. During the First World War the Germans cunningly distributed a leaflet to black American troops exhorting them to lay down their arms. It warned that they were being deceived into fighting 'for democracy' abroad when they had no rights at home (Mullen, 1973). Even during the Second World War, black soldiers in the American South were refused restaurant facilities that were available to German prisoners-of-war! The civil rights legislation of the 1960s did much to underwrite equality of treatment and opportunity for all but, today, although the overall position of black Americans has improved, it is still considerably below that of white Americans in terms of various socio-economic parameters such as income, education and unemployment (Farley, 1985).

The black population of Britain in 1945 was about 10 000 (less than that in the eighteenth century), but during the subsequent two decades there were large waves of immigration from India, Pakistan and the West Indies. The research projects carried out by Political and Economic Planning (PEP) unequivocally showed the marked extent of racial discrimination in Britain in sectors such as housing and employment. The PEP used West Indian, Asian and White actors to go out for job interviews. They also sent out similar application letters for an advertised job, one in the name of a white British applicant and the other in the name of an immigrant. A European foreigner (usually a Greek or Italian) was included to control for discrimination against foreignness and not colour. The results of the last two PEP surveys (Smith, 1977; Brown, 1984) – and these followed the promulgation of Race Relations Acts – demonstrated that West Indians and Asians consistently suffered discrimination more often than did Greeks or Italians. There was little difference in the discrimination experienced between West Indians and Asians. Thus skin colour, and not 'alien status', seemed to be the operative factor. As Arnold Toynbee wrote: 'The . . . (Negro) may have acquired the White man's culture and learnt to speak his language with the tongue of an angel . . . and yet it profits him nothing if he has not changed his skin.' In the late 1960s and the 1970s the emergence of Enoch Powell (with his 'rivers of blood' prophesies) and the National Front (with its 'Keep Britain White' slogans) stirred up British racism, and during the 1980s helped to precipitate the worst street rioting to be seen in English cities for more than a century. Lord Scarman cited 'racial disadvantage' as a major factor in the unrest, with its attendant bad housing, economic problems, unemployment and underprivilege.

Colour prejudice and racial discrimination are well documented in Australia where the Aborigines, who have all the political and civil rights that whites do, nevertheless exist in extremely poor conditions when measured against the prosperity of the rest of the country. They often live in poverty and misery, in overcrowded shanties on the outskirts of towns or under bushes and bridges. They are gaoled ten times more often than any other group (and deaths in detention are frequent); they have a higher rate of unemployment and chronic alcoholism, a shorter life expectancy and a higher infant mortality than other Australians.

But, in recent times, skin pigmentation has nowhere been more decisive in determining power, wealth and status than in South Africa. Since 1948, with the advent of the white Nationalist Government, the system of *apartheid* (separate development) has enshrined racial discrimination in the statute book. Separate facilities for whites and non-whites were decreed in all spheres of national life. The country became a true 'pigmentocracy' in that the franchise, and therefore the executive and legislative powers, resided strictly in the white domain. Apartheid has been buttressed by a cat's cradle of laws and regulations which have controlled virtually every aspect of human life, including residential areas and sexual intercourse across the colour line. In the earlier phase of apartheid, race classification boards were instituted to determine what racial label should be affixed to persons of doubtful ethnicity. Skin colour, although important, was only one of a checklist of physical characteristics used in the arbitration process. The grand design of apartheid also led to the establishment of different ethnic homelands for the South African blacks themselves.

The implementation of the apartheid policy was devastating in the misery and humiliation it wrought. It engendered feelings of violent bitterness and hostility between whites and people of colour, especially as the segregated amenities (e.g. in health care, education and housing) for non-whites were consistently inferior to those for whites. Unrest and violence became endemic to South Africa from the mid-1970s, and these were accompanied by the exercise of stringent repressive and emergency measures. During the 1980s the Government was forced to modify its ideological stance – mainly because of continuing internal and international pressure. The most offensive trappings of apartheid (such as segregation in public places, the miscegenation laws and the 'pass laws') were dismantled, but these so-called reforms were dismissed merely as cosmetic changes to the face of apartheid while its lineaments remained intact. It was only in 1990 that the government itself declared that the apartheid era was dead, and in so doing opened the way for the abolition of the structural racism of the past and the negotiation of a new political dispensation.

Passing for white

In any colour-conscious society where the possession of a white skin spells privilege, social acceptance and opportunity, some 'black' individuals of marginal pigmentation will make a bid for a crossing of the colour line. This process is known as 'passing for white' or 'playing white', and in the United States of America (particularly before the Civil Rights era) the practice was not uncommon in light-skinned people who had some degree of Negroid ancestry and were therefore regarded by the dominant society as black and socio-politically subordinate (the rule of hypodescent). The underlying philosophy was that 'if you can't beat them, then join them'.

Skin lightening creams were used as an aid to 'joining', sometimes combined with hair straightening and hair dyeing.

The reasons for passing have been complex and there are different types of passing. The most extreme variety involves a total alienation from the black group and an emotional identification with white society. This represents an identity change with a repudiation of all family and social networks, and such actions provoke great resentment and hostility from the parent community. There have even been instances of wealthy, light-skinned black American couples who have adopted white children rather than risking the birth of a dark-skinned heir who would expose their ancestry. The other motive for passing, and a more acceptable one, has been to achieve financial advantage by securing better employment. In this situation the individual would exploit a white skin for material gain without sacrificing family relationships or moving home.

In retracing South African social history, Fredrickson (1981) has noted that clear signs of colour prejudice and discrimination emerged only after the British took over the Cape Colony in 1795. Prior to that, non-white freedmen were included on the roll of 'free citizens' and there was a high degree of social acceptability for legal intermarriage – in fact, between 1688 and 1807 about one-quarter of the founding marriages of future Afrikaner families involved one spouse (usually the wife) who had some degree of non-white ancestry. The descendants of these intermarriages (called 'Cape Coloured' in the South African classification) were often sufficiently white to succeed in passing (see Table 7.1, note e). At the turn of the twentieth century in the Cape, the crossing-over of 'Cape Coloureds' was of such dimensions that it siphoned off a large proportion of what would have been a leadership class (Fredrickson, 1981). (A definition of a 'Cape Coloured' person was 'one who has failed to pass for white'!)

A survey of 'Coloured' subjects in Johannesburg in 1973 (Unterhalter, 1975) showed that just under half the respondents (and particularly those of higher socio-economic status) approved of passing for white: pragmatic considerations had transcended their allegiance to the 'Coloured' group. Interestingly, less than 10 per cent of respondents (and particularly those of lower socio-economic status) approved of South African Negroids passing for 'Coloured'. The majority had negative attitudes towards the Negroids and wanted to maintain social distance from them.

The black consciousness movement

Since the 1960s there has been a development which has undoubtedly had a distinct effect on colour perception. This is the *black consciousness* or *black identity* movement, the aims of which have been *inter alia* to abrogate all the negative concepts of blackness, to affirm the positive features and to assert a vigorous sense of pride and self-esteem among black people in being black. 'Black is beautiful' became a new slogan. In a sense, black consciousness was the opposite state to passing for white and the movement made considerable progress in eradicating past stereotypes by providing a diametrically opposed set of attitudes and values. Frantz Fanon (1967) has given a detailed psychological and philosophical analysis of the state of being black with its deep sense of inferiority and desire to be white. 'I propose nothing short of the liberation of the man of colour from himself', he wrote. Black consciousness aimed for just such a psychological liberation and it also forced white society to accept a new and positive definition of blacks.

Colour symbolism, with its almost universal attribution of 'good' adjectives to whiteness and 'bad' adjectives to blackness, has been discussed above. It is possible that the designation of racial groups by colour names may link the connotations of a particular colour to the racial group itself (Williams & Carter, 1967). In other words, the constant association of black with a racial concept like Negroid would invest the latter with the negative attributes of the former. This type of conditioning may therefore influence the way different racial groups are perceived, especially as black–white colour meanings are developed during the preschool years when racial awareness is beginning to emerge. Indeed, the pro-white/anti-black colour bias was found to be evident in both black and white American children from the age of 3 years onwards (Williams, Boswell & Best, 1975).

Another potent association that young white children learn to make with the word 'black' is its relationship to dirt ('Your hands are black – go and wash them', and, with the mission accomplished, 'They're lovely and clean now'). The idea of Negroids being dirty was actually captured by Victorian soap manufacturers who advertised their product by showing a black child's skin turning white after washing with their soap (Fig. 10.1). Then, the subtle impact of children's literature reinforced the negative colour symbolism of blackness during the formative years of life. Stories like *The Water Babies,* Dr Dolittle's adventures and *Little Black Sambo* abound with allusions and imagery which glorify whiteness and debase blackness (Isaacs, 1963; Milner, 1983).

During this century in North America there has been vehement argument about the terms used by Negroids to describe themselves (Isaacs, 1963). There had been numerous variations on the theme: blacks, negro (small 'n'), Negro, Coloured, Africo-American, American

Fig. 10.1 Victorian soap advertisement which associates the blackness of the African boy with dirt while the whiteness of the European girl is presented as an ideal of purity and beauty. (Courtesy of the Merlin Trading Company Limited, London.)

Negroes, Negro Americans, Aframericans. Current usage has favoured 'black', but the term 'Afro-American' is becoming popular. Although not provoking so much controversy, the term for 'white' people has also been problematic and the word 'Caucasian' has been criticized for its inappropriateness (Freedman, 1984).

The philosophy of black consciousness was that in order to destroy the prejudices and dogmas of the past, black culture, black history and black art would have to be re-created. *Negritude* was a literary compaign and a political faith inaugurated in the 1930s by Aimé Césaire in the Caribbean and Leopold Senghor in Africa. Negritude gave birth to a French literature which extolled blackness and ennobled African heritage. Not only was it a vehicle of revolt against the injustices of colonial rule and the conventional anti-black stereotypes but it tried to establish a new spirituality and value system for blacks, independent of the norms of the West. In the 1960s the cult of negritude had its parallel in other fields. Blacks demanded their own sociology and their own psychology, and they launched the appropriate organizations. *Pari passu* with the socio-cultural and intellectual components of black consciousness went the political arm, which proclaimed the idea of 'Black Power' and generated militant anti-white movements like the 'Black Panthers'. The beginnings of the black power credo were already in evidence with the disastrous failure of white liberals to become integrated within the civil rights movement in the American South in 1964 and 1965 (Pouissant & Ladner, 1968).

Several workers have investigated the effects of the black identity movement on the connotative meanings of black and white. A 1969 study of American college students found that for blacks (but not for whites) black had become more positive (good) and white less positive, the changes being most pronounced among the subjects most committed to black separatism (Williams, Tucker & Dunham, 1971). With regard to the pro-white/anti-black bias in children, there have been some conflicting findings. The American surveys have tended to show a reversal of the previous position (Ward & Brown, 1972; Teplin, 1976), whereas the studies of West Indian (Afro-Caribbean) and Asian British schoolchildren during the 1970s did not suggest any fundamental move away from pro-white/anti-black bias (Young & Bagley, 1979; Davey & Norburn, 1980; Milner, 1983). It is interesting, however, that the children of Jamaican parents in London had less pro-white bias than children of rural Jamaicans (despite the fact that the latter rarely met white people) and that black children from Nigeria and Ghana showed even less pro-white bias than did the British West Indian children (Young & Bagley, 1979).

An interesting phenomenon has emerged in the past few decades. Fair-skinned Caucasoids will spend hours in the sun in order to obtain an attractive tan (see Fig. 3.1), although at considerable risk of UV-induced skin damage. This quest for a tan has had its impact in other ethnic groups. Japanese men have begun to accept that a sun-tanned skin in a young woman may well impart a healthy attractiveness (Wagatsuma, 1968). To some extent the notion of 'whiteness' may be linked to paleness, and therefore to illness and death, while a light brownish complexion suggests fitness and well-being. There is also a class-related explanation applying to Caucasoids in Europe and North America. In the past ordinary people toiled in the fields and on the farms and they became progressively tanned from solar exposure: a pale skin then signified affluence and an aristocratic lifestyle free from hard manual labour. In the present era the tables have been turned: many Europeans and North Americans have indoor jobs, and it is now only the wealthy who can afford the leisure time outdoors and on holiday to acquire a tan.

The foregoing discussion has highlighted the socio-political implications of skin colour. It must be emphasized that racial differences in skin colour represent biological adaptations to different geographical regions, the evolutionary context of which is debated in the next chapter. Racism is the reaction to, and interpretation of, these differences by others and it has to be addressed within a sociological framework. Perceptions of colour differences in a society are changing and subject to factors such as the political milieu. Being black, for example, in a future democratic South Africa is likely to be evaluated more positively than it was under the apartheid regime. In the United States of America, the affirmative action programmes have induced some whites to pass for black in order to gain easier opportunities for promotion.

Sociological perspectives of pigmentary disorders
Albinism
Before albinism gained acceptance as a genetic biochemical disorder a great deal of curiosity surrounded the existence of albinos among Negroid populations, and they were even exploited commercially in the form of exhibits to the British and American public. The Cuna Amerindians, who have one of the highest prevalences of albinism in the world, regard their albinos as superhuman (Keeler, 1963) and prohibit them from marriage. This marriage handicap also applies to albinos among the Hopi Amerindians of Arizona and the Zuni and Jemez in New Mexico. In other respects these New World albinos are treated with kindness,

tolerance and even respect, although in the past the infanticide of albino newborns was widespread in Cuna society.

The attitudes of African Negroids to albinos have differed from country to country (Kromberg, 1985). Among the Fula people of the Sudan albinos lived on the charity of others, who gave generously to them to win favour from heaven. In Ghana and Dahomey albinos were believed to be under divine protection; in the Congo and Angola they were revered; in Kenya and Uganda they were perceived as curiosities and kept in the households of kings and great chiefs; and in Gabon they were deemed to be unlucky and were frequently killed. In Nigeria, even in contemporary times, albinos are regarded as repulsive and they are linked with various unfavourable superstitions. Every major language in that country had a name for them, usually a derogatory one which focused on their whiteness as a reminder of the colonial past (Okoro, 1975). In southern Africa infanticide of albinos was reported as being fairly frequent, and David Livingstone himself noted this practice during his extensive travels in Africa. There were several myths in Southern Africa relating to the origin, powers and fate of albinos, for example, that they had special spirit-powers and would not die naturally but just disappear when the time came (Kromberg, 1985).

Kromberg & Jenkins (1984) assessed 35 South African Negroid albinos and a normally pigmented matched control group. The former were as well adjusted as the controls and did not admit to any interpersonal or psychosocial problems, although they had a low marriage rate. The pigmented controls had reasonably positive attitudes towards the albinos but, surprisingly, they were more dissatisfied with their own physical appearance than were the albinos! It is of interest that these albinos did not identify with South African whites and that very few attempted to pass into a lighter-coloured ethnic group.

Kromberg, Zwane & Jenkins (1987) then evaluated the impact of the birth of an albino child on black South African mothers. The mothers were initially depressed, uncomfortable at being in close contact with their infants, and reluctant to hold and breast-feed them. This reduced level of interaction (compared with controls) lasted for three months, but the mothers still expressed feelings of unhappiness up to the time when their babies were nine months of age. Thus, the birth of an albino baby induced a weakening of maternal attachment in the early stages. The reasons for this are complex and may not have been due to the albinism but to the non-specific effect of rearing a baby with a congenital handicap. Furthermore, an albino baby's visual deficit may have impaired eye contact with resultant impoverishment of mother–child bonding. Fuller

& Geis (1985) have emphasized the potential significance of a situation where a newborn's skin colour varies greatly from that of the parents. Negative attitudes may ensue, such as feelings of rejection or doubts about paternity.

Vitiligo

The previous chapter noted that this disfiguring disorder was erroneously equated with leprosy. In southern India vitiligo is known as 'white leprosy' and affected individuals there (especially those whose disorder involves the face and exposed regions) have virtually no chance of employment or marriage.

In dark-skinned people vitiligo is a psychosocial disaster with devastating effects on self-esteem. Unfortunately, the disorder is grossly unsightly ('I look like someone threw a bottle of bleach at me', said one patient) and many sufferers develop neurotic symptoms – feelings of inferiority, resentment, shame, anger, anxiety and depression (Ortonne *et al.*, 1983). In an American questionnaire survey of 326 vitiligo patients (of whom 222 were Caucasoid and 90 Negroid), 57 per cent claimed that people stared at them, 20 per cent that they had been the victim of rude remarks, 23 per cent that the condition interfered with their relationships with the opposite sex, and 8 per cent that they had experienced job discrimination (Porter *et al.*, 1987). Some sufferers will spend fortunes in order to get treatment which, in any case, is difficult, prolonged and not always effective.

Stolar (1963) was one of the pioneers in the depigmentation treatment of extensive vitiligo. In this treatment monobenzyl ether of hydroquinone (see p. 161) is applied as an ointment to the remaining islands of *normal* skin to bring about a homogeneous white coloration of the whole body. Depigmentation becomes evident after 4–8 weeks of treatment, but it requires about 2–3 years to effect total depigmentation which, in patients with vitiligo, remains permanent thereafter. This treatment effectively turns Negroid patients white (hair and eye colour are unchanged), a process that in itself produces major psychological effects on the individuals and their families. Although the duration of therapy is sufficiently long to allow for gradual adaptation, interviews with some patients who had undergone the pigmentary conversion (Irons, 1971) revealed that none was particularly happy about their new look – except it was undoubtedly preferable to the horror of vitiligo. However, Stolar was besieged by requests from normally pigmented blacks for a black-to-white transformation to lift them on to a higher social and economic plane (Irons, 1971).

Nelsons's syndrome

This is a condition (see p. 125) that can produce such a marked increase in skin and hair pigmentation that affected white subjects are identified as being coloured (Fig. 8.2). From time to time newspaper articles give considerable publicity to the experiences of particular patients with Nelson's syndrome. These patients have described the aversive responses of their (white) community to the colour change: they have been shunned by erstwhile friends and acquaintances who were too embarrassed to be seen fraternizing with a dark-skinned person. In a report of Nelson's syndrome from Lancashire, England (Buckler *et al.*, 1988), the patient's wife left him because she could not accept his heavy pigmentation.

Two cases of the syndrome from South Africa have highlighted the problems specific to that country. One 17-year-old Transvaal girl contracted the disorder as a teenager and became so hyperpigmented that she was regarded as a 'Coloured' (Schumacher, 1973). In her apartheid society she became *persona non grata*; she was treated with suspicion and hostility by whites and taunted with racist remarks at school. The ordeal precipitated her into a state of depression and anxiety. Another patient, a white woman aged 40 years, underwent similar tribulations as her skin colour turned progressively darker. She was thrown off 'whites-only' buses and had to obtain a special card from the transport company to prove to busdrivers and conductors that she was actually white! When she moved to a different suburb in Cape Town, she received a petition from the neighbours accusing her of trying for white and stating that she was not welcome. Her husband and family deserted her because they could not endure the barrage of insults and humiliation. Her mother refused to let her attend her father's funeral because 'I don't want to be embarrassed by your black body at daddy's grave' (Joseph & Godson, 1988). A sharp contrast in social attitude and perception was the example of a 23-year-old white Brazilian woman with Nelson's syndrome. She enjoyed the change in pigmentation from white to dark because fellow-Brazilians (including young men) stopped her in the street to admire her new colour and particularly its combination with her long, straight hair (Anonymous, 1973).

Chemical skin darkening

All the above descriptions relate to abnormal hyperpigmentation caused by disease, but there are also instances where adventurous investigators have turned themselves black by artificial (chemical) means. The project is accomplished by the ingestion of psoralens – photosensitizing drugs stimulating melanin production (see p. 53) – coupled with regular and prolonged sun exposure. This regimen results in a remarkable degree of skin darkening, which can be supplemented with hair dyes, black wigs and black contact lenses to complete the conversion.

The first such exercise was undertaken by John Howard Griffin in the southern United States and it is faithfully recorded in his book, *Black Like Me* (1960). About a decade later, Grace Halsell repeated the experiment and chronicled her trials and hardships as a 'black woman' in Harlem and Mississippi in her book, *Soul Sister* (1969). In both of these sagas the theme of racism permeated almost every situation the authors encountered. The value of these particular commentaries is that they portray the reaction that *whites* have to being black. James Baldwin once asserted that no white person, not even of an oppressed minority, could know what it was like to be black in a white racist society. But Griffin and Halsell gained first-hand and agonizing insights into what happened (in the 1960s) when, in Griffin's words, 'a so-called first-class citizen is cast on the junk-heap of second-class citizenship'. And, in concluding her book, Halsell bluntly states that she could never repeat the venture 'because now I know what it cost me, psychologically, to bear, for one minute in time, what every black American bears all his life: discrimination, segregation, injustice'.

The conclusion that crystallizes from the bold and harrowing exploits of Griffin and Halsell and from the testimony of individuals with Nelson's syndrome is that skin pigmentation – and skin pigmentation alone – was the sole criterion for pronouncing on the measure and worth of human beings. Intelligence, personality, courage, cultural attainments or speaking accent had not a whit of influence if the skin melanin content exceeded a certain threshold. MacCrone (1957) put it well: 'The black man wears a livery that he cannot discard, though he may have discarded everything else that could possibly be regarded as a justification or excuse for his exclusion.'

Skin colour and blood pressure

Various studies in the United States of America have compared the blood pressures in blacks and whites. Not only do blacks have significantly higher pressures but they are also more predisposed towards hypertension (abnormally high blood pressure) and its complications than are whites (*Lancet*, 1980). Several biochemical and physiological explanations have been proposed – common denominators between the catecholamines (which increase blood pressure) and melanin – but there are also important sociological considerations, especially as hypertension is a stress-related disease.

A study of blood pressure levels in black and white American males in four racially segregated areas of Detroit found that the pressure was highest in black males in the 'high-stress' area (i.e. one that was over-crowded, poorer, less stable and more crime-prone) and that, among black subjects, skin colour was positively correlated with blood pressure (Harburg *et al.*, 1973, 1978a). Black males in the 'high-stress' area were therefore darker than middle-class blacks in the 'low-stress' area. Suppressed hostility emerged as a clear-cut contributor to elevated pressure, and this suggested that black Americans with the darkest skins had suffered the greatest discrimination and emotional stress. In the whites, on the other hand, the skin colour relationship was reversed: the darker the individual, the lower the pressure – those of Mediterranean extraction had the lowest pressures and those from northern Europe the highest (Harburg *et al.*, 1978b). Among black American females in Charleston, South Carolina, skin colour (measured by reflectance spectrophotometry on the unexposed skin) did not have a direct association with blood pressure (Keil *et al.*, 1981). The latter was more influenced by the educational level of the subjects, a higher pressure accompanying a lower educational level.

In Puerto Rico, dark-skinned men had a lower socio-economic status and slightly higher blood pressures than light-skinned men (Costas *et al.*, 1981). Skin colour still showed a small but statistically significant association with blood pressure after the effect of socio-economic status had been removed.

In South Africa, urban blacks have a hypertension rate similar to that of American blacks. South African blacks (urban Zulus), whites and Asians (Indians) have hypertension prevalences of 25 per cent, 22.8 per cent and 19 per cent respectively (Seedat & Seedat, 1982). This ranking is not directly related to skin colour, as otherwise the Asians would have occupied an intermediate position. Furthermore, rural blacks (Zulus) had a prevalence rate of hypertension of 10.5 per cent, less than half that of their urban counterparts (Seedat, Hackland & Mpontshane, 1981). Further investigation disclosed a connexion between hypertension and social stress factors (as experienced by the urban Zulu) – acculturation, detribalization and exposure to political ideology (Seedat, Seedat & Hackland, 1982).

In conclusion, skin colour is not directly implicated in the causation of hypertension. Although there might appear to be an association in terms of black–white differences in the disorder in the United States of America and in South Africa, it is not skin colour *per se* that is linked with hypertension but rather the psychosocial correlates of skin colour within a particular society. Thus, having a black skin in Detroit is more pathogenic (*vis-à-vis* hypertension) than having a white skin in Detroit, which is more pathogenic than having a black skin in a rural South African setting.

Notes

1 James (1981–2. *New Community,* **9**, 19–30) believes that the offensive meanings of 'black' developed during the medieval and Renaissance period (from about AD 1200 onwards) and that these sprung from ideas of spiritual light and darkness (rather than physical light and darkness) which were introduced by the liturgy and visual symbolism of the Church.

2 Certain countries in Africa (e.g. Senegal) have prohibited the import of skin lightening compounds. In 1975 the then President of Uganda, Idi Amin, banned skin lighteners because he claimed that they made women look unnatural, and that the practice was acquired from former colonial masters who believed that one had to be white to be civilized.

3 The high frequency of albinism among the Hopi Amerindians has been explained by the fact that, as albino males had to be kept indoors during daylight hours because of their aversion to the sun, they had ample opportunity to engage in sexual activity, thereby propagating the albino gene (Woolf, C. M. & Dukepoo, F. C. (1969). *Science,* **164**, 30–7).

4 The term 'blue blood', which denotes high or noble birth, is of Spanish derivation and it conveys the fact that the pure-blooded Spanish aristocracy had no Moorish or other ethnic admixture. Their skins were so fair that the underlying veins were visible.

5 The irony of 'Aryans' being linked with Nordic characteristics (tallness, blondness, etc.) was that the most important members of the Nazi government – Hitler, Hess, Goering and Goebbels – had appearances which were markedly discrepant with the Nordic ideal, a situation that caused considerable embarrassment to the Party. Furthermore, the fact that so-called 'Aryans' and Jews were indistinguishable physically was as much as admitted by the Nazis by their act of decreeing that all Jews had to wear a yellow badge (Star of David) to identify themselves as Jews.

11 *The evolution of skin colour*

Perhaps the most notable feature about skin colour is its geographical distribution, and indeed there are skin colour maps (the best known one by the Italian geographer, Biasutti) which plot the varying grades of skin pigmentation among indigenous populations throughout the World. Examination of such a map (Fig. 11.1) does give a general impression of a skin colour 'cline', i.e. a pigmentation gradient from high to low latitudes. In very general terms, the lightest skins occur in the northern latitudes and, as one progresses in a southward direction, so colour gradually deepens. In Europe the fair-skinned Scandinavians in the north-west merge into the swarthier Mediterraneans in the south; in North Africa and the Middle East olive complexions predominate, while a much deeper pigmentation characterizes the inhabitants of tropical Africa. Indeed, some of the blackest skins occur on the African continent in the belts between latitudes 10° and 20° (N and S). In Asia there is a similar picture with lighter-skinned people in the north, brunet coloration in

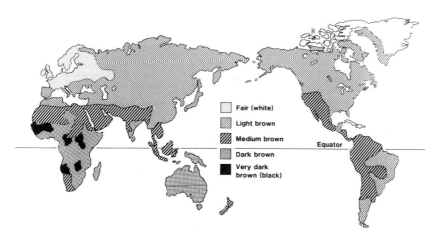

Fig. 11.1 Example of skin colour map (unexposed skin) of indigenous world populations based on Italian geographer, Biasutti. Overall pigmentation gradient is shown here, skin colour progressively darkening with decreasing latitude.

187

south-west Asia (including much of India), and dark skins in the south of India, Melanesia and Australia. In Australia the Aborigines of the northern regions (i.e. Arnheim Land) are more pigmented than those in south-east Australia.

The skin colour gradient in the New World is much less marked or consistent, probably because of the relative recency of immigration into North and South America. But even here skin colour maps show some variation, with the lightest peoples in the northernmost parts of North America and in the southernmost parts of South America.

The skin colour map of Biasutti was based on data derived from the von Luschan tablets (see p. 98). Not only did the latter method of measurement have definite shortcomings but, for regions where no information existed, Biasutti simply filled in the map by extrapolation from findings obtained in other areas! These older maps must therefore be used with circumspection in assessing clinal gradations of skin colour. The advent of portable reflectance spectrophotometry (see pp. 99–103) has provided anthropologists with an objective, reproducible and scientifically acceptable means of quantifying skin colour, and the construction of a skin colour map using readings from this technique (Tasa, Murray & Boughton, 1985) is a marked improvement on earlier attempts (Fig. 11.2).

Roberts & Kahlon (1976) assembled data from world-wide reflectance studies and correlated these with such environmental variables as latitude, temperature, humidity and altitude. They found that latitude *per se*

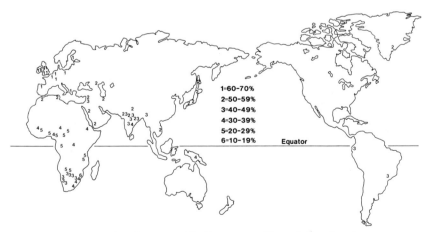

Fig. 11.2 Skin colour map of indigenous world populations based on reflectances at 685 nm (Filter 609) at inner upper arm, measured with EEL reflectance spectrophotometer. Clinal distribution (latitude) of skin colour is evident. (After Tasa *et al.*, 1985, with modifications.)

accounted for a very great proportion of the total variance in skin reflectances of human populations. The next most important factor was temperature, although as an independent variable it had a much smaller effect than latitude.

The close association between latitude and skin colour (see Fig. 11.2) was confirmed in a recent and more comprehensive survey (Tasa *et al.*, 1985) which found a positive correlation of 0.82 between skin reflectance and distance from the equator. Because latitude itself (and not temperature) was the major component in the geographical distribution of skin colour, ultraviolet radiation (UV) was presumed to be the most crucial determining factor (Roberts & Kahlon, 1976). The amount of UV received at the earth's surface (see pp. 45–8) is inversely related to latitude – the higher the latitude, the lower the UV. Put simply, it seems that from a geographical standpoint the darker the skin pigmentation the greater is the ambient UV.

Ultraviolet radiation and melanin

The photoprotective properties of melanin have been discussed (see pp. 59–63). The principal factor protecting Negroids from the deleterious effects of UV-B is their high content of epidermal melanin. If a black skin evolved in the tropics then this supposes that those with a 'white' skin would have been eliminated from these latitudes by natural selection. It is therefore necessary to examine those effects of UV-B which may potentially imperil the survival of relatively non-pigmented people.

Skin cancer is far commoner in fair-skinned Caucasoids than in Negroids and the onset of the disease is directly related to the cumulative amount of UV exposure. Was susceptibility to skin cancer the selective force against light pigmentation in the tropics? Blum (1961) has contended that, because skin cancers (except for the rare malignant melanoma) are usually not lethal and tend to appear only in later life (when the reproductive period has passed), their occurrence could not have significantly influenced survival. This argument is not very convincing – in Nigeria and Tanzania all the Negroid albinos studied exhibited skin cancers or premalignant lesions by the age of 20 years (Okoro, 1975; Luande *et al.*, 1985), and a mere 6 per cent of the Nigerian albinos were in the age range 31–60 years compared with 20 per cent of the non-albinos (Okoro, 1975). Early African hominids spent a great deal of their daylight hours in gathering and hunting, and prolonged exposure to the strong tropical UV would have wreaked severe damage on unpigmented skin with the emergence, as in African albinos today (see Fig. 9.4), of skin cancer – and probable death – during early reproductive life.

Whereas skin cancer is a chronic sequela of UV exposure, sunburn is an acute effect. Sunburn is produced by the UV-B portion (280–320 nm) of the UV spectrum (see pp. 43–4). It manifests as a redness of the skin which in severe cases can progress to tenderness, pronounced swelling, and blistering: these signs may be accompanied by chills, fever, nausea and even delirium. Sunburn is a painful and debilitating reaction, as any reckless sunbather will know. Melanin is an excellent sunscreen, and dark-skinned Negroids are practically immune to sunburn. An adaptive response to sunburn is the formation of a tan which, as it develops, should increasingly protect against further sunburn. But some light-skinned Caucasoids (skin types I and II) have little or no tanning capacity (see Table 3.1) and continue to get sunburnt on re-exposure whereas the indigenous Greenlander, for example, has a remarkably high index of tanning (Ducros *et al.*, 1975).

The critical question, however, is whether sunburn in itself was a sufficiently potent factor to have eliminated light-skinned individuals from the tropics. Blum (1961) minimized the importance of sunburn in an evolutionary context: it was an acute state which did not cause long-term damage and it was seldom disfiguring enough to affect sexual selection. Blum's contention does not appear to be valid.

The most important activity of Pleistocene hominids in the African savannahs was gathering and hunting.[1] Recently there has been a re-evaluation of the food resources of early *Homo* and it is estimated that at least half (and possibly two-thirds or more) of the diet derived from the gathering of plant and vegetable foods. Although current thinking dictates no hard-and-fast pattern about sex roles in primitive societies (Dahlberg, 1981), women are more likely to perform the gathering operations, partly because child-rearing and child-carrying restrict their mobility and manoeuvrability.

In view of the starkness of early hominid subsistence, it is hardly conceivable that persons of fair, sunburn-prone complexion would have been able to endure lengthy daily exposures to the tropical savannah sun. Sunburn, with its pain and discomfort, would have seriously incapacitated their gatherer–hunter activities. Bortz (1985) has formed the hypothesis that physical exercise was an integral part of *Homo*'s history, especially the human ability to run long distances in the heat – far longer than virtually any other animal – in pursuit of prey ('chase' hunting). As Bortz remarked: 'But no matter how the hunter hunted, the best hunter, the alpha runner, had survival and reproductive advantage'. An inferior hunter, such as one disabled by sunburn, would not only fail to keep himself in food but would lose standing within the social group. The ethnographic literature has highlighted the widespread practice of males trading meat in exchange for copulation (Hill, 1982): good hunters obtain more wives than poor hunters, sire more children and thereby contribute disproportionately to the succeeding generation.[2]

Humans, with their supreme capacity for sweating (strongly promoted by hairlessness), rapidly lose heat during prolonged physical exertion in hot climates, giving them a great cooling advantage over other animals. Although humans need regular supplies of water to offset dehydration human sweating is an essential thermoregulatory mechanism for preventing excessive elevations in body temperature (hyperthermia). Sunburn would probably have its most profound impact on survival by its interference with sweating. UV produces a diminution in the sweating rate of Caucasoid, but not Negroid, skin by blockage of sweat gland ducts (due to epidermal damage) and by a possible reduction of the sweat secretion itself (Thomson, 1951). The failure of a light-skinned person with sunburn to sweat efficiently under a solar heat load and during physical exertion would lead to hyperthermia and death.

Many of the evolutionary hypotheses of skin colour focus on selective factors in the adult, but perhaps a more realistic interface would be at the infant level. The role of the woman as gatherer did not exempt her from childcaring and, while she foraged for food, she would carry and feed her baby. The child would thus be exposed to the full force of the tropical sun with its high quota of UV-B. The skin of babies is thin and less pigmented than that of older children and adults. Newborns in New Guinea have relatively little melanin at birth but they acquire the full adult complement by the age of 6 months (Walsh, 1964). Thus, those reasons that account for a light-skinned adult's vulnerability to solar radiation would apply *a fortiori* to the infant. With a thinner epidermis than the adult and less melanin, a relatively unpigmented baby in the equatorial latitudes of Pleistocene Africa would have undergone a swift demise from extensive sunburn, hyperthermia and dehydration. 'White' skins would therefore have been selected out by natural selection very early in the human lifespan. Moreover, the level of skin pigmentation in infancy is the crucial criterion in terms of UV-B. The subsequent darkening and the skin colour achieved in adulthood may be less important in evolutionary terms.

Skin pigmentary differences between Negroids and Caucasoids only reveal themselves at 32 weeks of gestation (Post *et al.*, 1976): before 32 weeks the foetuses of Negroids and Caucasoids are indistinguishable by reflectance spectrophotometry. It has also been a curious finding that at birth Negroids (Post *et al.*, 1976) as well as New Guinea infants (Walsh,

1964) had lower skin reflectances (i.e. higher skin melanin concentrations) at the forehead than at the inner aspect of the upper arm. This was puzzling because it was always assumed that the forehead was darker only because of its habitual exposure to sunlight. A plausible evolutionary explanation for the relative darkening of the face and forehead at birth is that sunburn of the newborn's face and lips would jeopardize sucking behaviour and impair breast-feeding. The facial region therefore needs added melanin protection right from the time of delivery. Similarly, the specific pigmentation of the breasts (areolae) during pregnancy (Garn & French, 1963) may be an insurance against sunburn of the nipples which, during the breast-feeding phase, could have life-threatening implications for the baby. And the high melanocyte density and pigmentation of the genitals, even at birth (see pp. 6–7), may safeguard these organs from the peril of UV injury and preserve reproductive capacity.

Thus sunburn constitutes a first-line selection pressure in the evolution of dark skin colour. If it is assumed that early hominids arose in the African savannahs and subsequently migrated northwards, then they must have become progressively depigmented as they penetrated the higher latitudes.

The likely reasons for this will be discussed shortly. But the inhabitants of latitudes above 40° N in Europe are fair-skinned whereas at the equivalent latitudes in Asia the inhabitants (Mongoloids) are darker. This difference may relate to two main factors. First, the climate in Asia is generally drier for any given latitude than in Europe and this would enhance the incident UV owing to a relative lack of cloud cover and mist. Second, Asia has the greatest average height of any continent except Antarctica; it is over three times higher than Europe and it is therefore exposed to more intense UV (see pp. 46–7). The higher UV influx in Asia would have exerted a selective effect on skin colour, even in far northerly latitudes. The exceptional tanning ability of the Inuit (Eskimos) (Ducros *et al.*, 1975) is possibly an adaptation to the high UV reflectivity of the ice and snow which blanket the Arctic during much of the year.

Heat load

In order to gain a better understanding of the arguments put forward here, the reader is referred back to Chapter 4 (pp. 63–8). It is a physical property of black objects to absorb heat, and Negroids absorb about 30–40 per cent more solar heat than do Caucasoids. Blum (1961) has maintained that the possession of a dark skin should be a disadvantage to the Negroid in a hot climate – the greater heat absorption and heat load would require more sweating (to prevent hyperthermia) and more water loss. Unless the latter can be adequately replaced, the person would succumb to dehydration and circulatory collapse. 'Thus', wrote Blum, 'if a Negro and a white-skinned man walked side by side, naked, in direct sunlight, across a desert where the temperature of the air was above that of the body, the Negro might be expected – other things being equal – to collapse before the white man'. Indeed, Blum facetiously suggested that Negroid skin would be best adapted to life on the snow-capped mountains near the equator, thereby capitalizing on both the heat-absorbing and the photoprotective properties of melanin.

Baker (1958) found that when Negroid and Caucasoid soldiers walked naked in the hot desert sun the Negroid group had the poorer tolerance. This is apparently the only study of its kind and it remains to be replicated. However, there have been numerous investigations of Negroid and Caucasoid heat tolerances in hot-humid heat (see pp. 65–6). All of these have demonstrated that, irrespective of body size or physique, Caucasoids are *not* superior to Negroids or to other pigmented ethnic groups: in fact, they have the highest rates of sweating and they may actually be inferior.

The use of the desert model by Blum is inappropriate because the stamping ground of early hominids was not desert territory. The Sahara has been a desert for only part of its history – 5000 years ago it was open grassland. Hominids appear to have evolved within the matrix of the tropical African savannah where summers tended to be hot and humid. Under these climatic conditions a black skin would have fared no worse in temperature regulation than a white skin and, if anything, it may have had a slight advantage because body water is better conserved by the lowered sweat production in Negroids.

Concealment

Camouflage plays a vital role in the animal kingdom for predator and prey alike (see p. 68). Cowles (1959) has proposed that the early hominids, weaponless and fireless as they were in the beginning, would have had to rely on concealment both for hunting success and for protection against predators such as leopards. A black skin would have been advantageous, especially in heavily forested areas or during times of low illumination (dusk or dawn). Cowles cites the fact that, in the jungle warfare of the Second World War, the Japanese reduced their visibility by darkening their bodies. Hamilton (1973) has disputed Cowles' hypothesis on the ground that black is not a particularly good colour for concealment. Humans would be more effectively camouflaged if they were to match the background. Forest animals such as antelopes or monkeys are generally brown or grey, not black, and the forest-dwelling Pygmies are lighter in colour than the adjacent Negroid villagers. Modern jungle and bush fighters wear a *patterned* green and brown garment which blends better both with the landscape and with other animals.[3] Hamilton also argues that, as humans are relatively large creatures, they should not be uniformly coloured black if they are to depend upon natural camouflage. Hamilton's arguments are cogent, especially as the bipedalism of the early hominids had propelled them out of the forests and into the open African savannahs where the colour black would have rendered them unduly visible. Similarly, in northern Europe, a light skin would have had camouflage value only in the snow-covered regions during the winter and early spring; elsewhere, and especially in summer, a whitish skin colour would have been visible against the expansive woodlands. Thus, the role of camouflage in the evolution of human skin colour does not appear important and it can probably be discounted.

Related to the subject of camouflage is that of colour vision. Post (1962) has noted that colour blindness has the lowest rates of occurrence in hunter–gatherer populations (e.g. Australoids, Inuits, Fiji Islanders and Brazilian Amerindians) and the highest rates in Europeans and Asians (especially Brahmans) who have long had settled community habitats. In explanation he proposed that colour blindness would be a severe handicap to hunter–gatherers in their quest for food and prey whereas it would be far less disadvantageous for the more settled pastoral–agricultural societies.

Resistance against disease

Wassermann (1965b) has developed an unusual hypothesis which relegates skin pigmentation to a secondary status. The primary evolutionary event was the successful adaptation of the Negroid to tropical conditions and his resistance to bacterial, parasitic and viral infections.[4] The principal organ of defence in humans is the reticuloendothelial system (RES), and the activity of the RES is inversely related to secretions of the adrenal cortex (corticosteroids). Corticosteroids suppress RES activity and this suppression makes the body more susceptible to infections; decreased corticosteroids allow for increased RES activity and therefore greater resistance to infections and parasitic diseases. Wassermann cites evidence that Negroids have reduced adrenocortical function which would in turn stimulate the RES and thereby prevent tropical diseases such as malaria. At the same time the relative adrenocortical deficiency would initiate a greater pituitary secretion of adrenocorticotrophic hormone (ACTH) and melanocyte-stimulating hormone (MSH) with a resultant increase in skin pigmentation. According to Wassermann, this scheme is the basis for the skin colour of the Negroid – it is tropical *disease* rather than tropical climate which is the selective factor leading to ethnic pigmentation.

There are certain problems with Wassermann's formulations. First, the concept of a human beta-MSH was abandoned subsequent to the publication of his hypothesis (see p. 33). But even if the 'pigmentary' pituitary hormones (ACTH and beta-lipotropic hormone) do have melanocyte-stimulating activity, their contribution to normal human skin pigmentation is dubious. In the days when plasma levels of MSH were measured in human subjects, there were no differences between Negroids and Caucasoids. Moreover, the loss of pituitary function in Negroids has no effect on their skin pigmentation.

Second, tropical diseases like malaria only started to flourish with the advent of agriculture in Africa (about 5000 years ago). It was the clearing of the land for cultivation purposes that created breeding sites for mosquitoes. Infectious diseases are generally spread with the formation of settled communities, where factors like overcrowding and the sharing of food and sanitary arrangements facilitate the exchange of bacteria and parasites. Thus, Wassermann's hypothesis of the primary role of disease in skin colour determination falls away. The greatest part of human evolutionary history occurred in the context of hunter–gatherer and nomadic societies where population density was low and infectious or parasitic disease probably non-existent.

Third, even in terms of Wassermann's proposal, there are inconsistencies. While Negroids are relatively resistant to tropical diseases such as malaria and yellow fever, they are highly susceptible – and more so than fair-skinned Europeans – to diseases like tuberculosis and cholera (Kiple & King, 1981). Negroids do not exhibit a general disease resistance due to increased RES activity but rather a specific immunity to tropical diseases based on centuries (or millennia) of exposure.

Wassermann's hypothesis, interesting as it may be, is not supported by the facts and cannot be taken seriously in the absence of further information and explanation.

Skin depigmentation: the vitamin D hypothesis

An important hypothesis has been advanced to account for the presence of light-skinned people in the northern latitudes of Europe. Before this hypothesis can be discussed and critically examined, a brief synopsis of the physiology of vitamin D is given below, with special reference to its role in preventing rickets.

Vitamin D

Vitamin D in the body is derived both from the skin (by far the more important source) and the diet. In the skin vitamin D is synthesized by the specific action of UV-B (wavelengths 280–320 nm), the optimum wavelengths lying between 295 and 300 nm (MacLaughlin, Anderson & Holick, 1982). The formation of vitamin D begins with a precursor substance, 7-dehydrocholesterol (7-DHC). The 7-DHC occurs in the epidermis and, to a lesser extent, in the dermis; it is chiefly the epidermal component which is converted into vitamin D. Within the epidermis (see Fig. 1.1(*b*)) over 90 per cent of 7-DHC is located in the Malpighian layer (i.e. basal layer plus stratum spinosum), the stratum corneum itself containing a negligible amount (Reinertson & Wheatley, 1959; Holick *et al.*, 1980). Negroid and Caucasoid epidermis do not differ in the quantity of 7-DHC (Reinertson & Wheatley, 1959).

When skin is exposed to UV-B, there is a photochemical (non-enzymatic) conversion of the epidermal 7-DHC into previtamin D; the latter is gradually transformed into vitamin D by a temperature-dependent process over 2–3 days (Holick *et al.*, 1980). This vitamin D then slowly diffuses into the blood vessels of the dermis, binds to a specific vitamin D-binding protein (an alpha-globulin) and is carried in the circulation to the tissues. Some of it is taken up by the liver and converted into 25-hydroxyvitamin D (25-OHD), which is one of the principal circulating metabolites of vitamin D: in fact, the blood concentration of 25-OHD is regarded as the best index of the vitamin D supply to the body (from both cutaneous and dietary sources). The 25-OHD is further metabolized in the kidney to 1,25-dihydroxyvitamin D (1,25-$(OH)_2D$) and it is this metabolite which is generally believed to be most important and biologically the most active form of vitamin D. The main actions of vitamin D (mediated predominantly by 1,25-$(OH)_2D$) are to stimulate the absorption of calcium from the gastrointestinal tract and to promote the normal calcification (mineralization) of bone. Except in abnormal states (e.g. vitamin D deficiency) the concentration of 1,25-$(OH)_2D$ is rigidly controlled and, unlike that of 25-OHD, it is not influenced by UV-B exposure or diet.

Vitamin D is also present in certain foods, the highest content being in fish liver oils and in fatty fish (e.g. herring, mackerel, salmon). Other items such as eggs and butter contain far less of it, milk (cow and human) is an inadequate source while meat, fruit and vegetables are practically devoid of vitamin D. For the treatment and prevention of vitamin D deficiency there are high-dosage preparations of the vitamin. However, over-ingestion of these preparations may produce vitamin D toxicity which causes raised blood levels of calcium as well as calcium deposition in various organs, especially the kidney (with the possibility of kidney failure and death).

Over-exposure to sunlight is not associated with vitamin D intoxication, and light-skinned Caucasoids in the tropics need not fear this complication (they can confine their worries to sunburn and skin cancer!). It has been neatly demonstrated by Holick, MacLoughlin & Doppelt (1981) that, during prolonged irradiation with UV-B, the synthesis of previtamin D reaches a plateau at a level equal to about 10–15 per cent of the original 7-DHC content, after which previtamin D is converted by the light into two inert compounds (lumisterol and tachysterol). This ceiling effect on previtamin D synthesis protects against UV-induced vitamin D toxicity, and there appears to be a further mechanism in which sunlight degrades skin vitamin D itself into a variety of photoproducts (Holick, Smith & Pincus, 1987; Webb, De Costa & Holick, 1989).

Because of the direct link between vitamin D production and sunlight exposure, it is not surprising that there is a seasonal variation in the blood levels of vitamin D (particularly at the higher latitudes) in accordance with the amount and intensity of solar radiation. Blood levels of 25-OHD attain a maximum in late summer/early autumn and a minimum during winter. Moreover, individuals who work outdoors show higher blood levels than indoor workers: this is observed throughout the year although both groups show the same pattern of seasonal fluctuation (Neer, 1985). There is also the altitude factor: places at higher altitude receive more UV-B, and the citizens of Detroit, Seattle and Boston (all at 42°–47° N) had comparable blood 25-OHD values, whereas inhabitants of Denver (nearly the same latitude (40° N) but a mile higher in altitude) had levels about 50 per cent greater (Neer, 1985). Conventional window glass absorbs all UV-B and the usual indoor (artificial) lighting emits little or no UV-B. Factors like cloud cover and atmospheric pollution (smoke, dust) impede the delivery of UV-B to the earth's surface and may therefore reduce cutaneous vitamin D synthesis.

Rickets

A deficiency of vitamin D in infants and children causes rickets; in adults (in whom bone growth has ceased) the condition is known as *osteomalacia*. In rickets there is defect in the calcification of growing bone so that

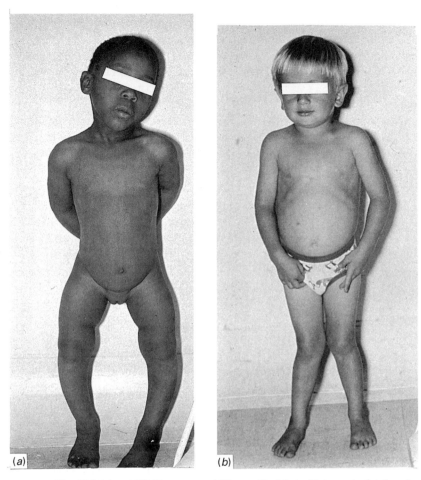

Fig. 11.3 (*a*) and (*b*) Two young children with rickets. Note marked deformity of the lower limbs which would have drastically impaired the hunting ability of early humans. (Courtesy of Professor François Bonnici).

the bones become structurally weak (especially during active growth periods) and are unable to withstand the various mechanical pressures placed on them. They consequently become deformed, and in severe cases there is a bending or twisting of the long bones with the characteristic bow-legs and knock-knees (Fig. 11.3). Fractures may occur, growth rate is stunted, and there may be distortion of the pelvis serious enough to make childbearing in later life dangerous and, in the absence of Caesarean section, potentially fatal for the baby and even the mother.

Rickets became increasingly prevalent in Britain and Europe during the industrial revolution with the emergence of smoke-polluted, over-crowded cities. The smog, together with dark and dingy tenements, deprived children of natural sunlight and the opportunity to synthesize vitamin D. Added to this was the notorious window tax, enforced by the British Parliament during the eighteenth century, which resulted in people bricking in the windows of their houses. Rickets became so common that by 1900 up to 90 per cent of children living in the major cities of Europe and North America suffered from it, although it was virtually absent in the rural areas.

The cause of rickets was finally confirmed after World War I when it was found that the administration of cod-liver oil, exposure to sunlight, or UV irradiation (with the mercury-vapour quartz lamp) all cured rickets whereas other forms of treatment (e.g. improved hygiene) were useless. These observations indicated that rickets was due to a deficiency of a fat-soluble factor which occurred in cod-liver oil and was also produced in the skin by exposure to UV.

Rickets has been largely eliminated as a public health problem in northern countries by the fortification of food with vitamin D (e.g. milk in the United States of America and margarine in Britain and Scandinavia), but two groups are still at risk. The first is the elderly population in whom low blood levels of 25-OHD are common. This is because the aged have reduced exposure to sunlight (from being housebound or institution-alized), reduced dietary vitamin D, and, as recently found, a decreased capacity to produce vitamin D from UV exposure (Holick *et al.*, 1987). The result is that the elderly are prone to an adult form of rickets (osteomalacia) which is associated with an increased risk of hip fractures. Asian residents in Britain are the second group of subjects who are still vulnerable to rickets and osteomalacia (Holmes *et al.*, 1973; Goel *et al.*, 1976). Since 1961 these diseases have been reported throughout Britain in Indian and Pakistani immigrants and particularly in their children, up to one-third of whom manifest clinical rickets. The most probable cause is inadequate exposure to sunlight but factors such as vegetarianism, the consumption of chapati flour, and genetic susceptibility may also play a part.

The occurrence of rickets and osteomalacia in sunny climates is usually observed in communities where cultural or religious norms dictate the wearing of heavy clothing or the avoidance of open air (e.g. the Bedouin women of the Negev desert in Israel). Negroid babies in Africa still experience rickets in the first year or two of life because they are breast fed, extensively swaddled and kept out of the sun. Sunscreens (see pp.

51–2) also block the epidermal absorption of UV-B. Topical application of the sunscreen, para-aminobenzoic acid, not only prevents the expected rise in blood vitamin D concentration after exposure to one minimal erythema dose of UV radiation (Matsuoka *et al.*, 1987) but, after prolonged use, reduces serum 25-OHD levels, even to the point of vitamin D deficiency (Matsuoka *et al.*, 1988). Thus, the elderly and others predisposed to vitamin D insufficiency should avoid the habitual use of sunscreens before going outdoors.

Vitamin D and evolution

Murray (1934) had put forward an ingenious hypothesis that attributed the evolution of a white skin to vitamin D synthesis. More recently Loomis (1967) has revived, refined and popularized this hypothesis so that it is now widely embraced in evolutionary thinking. The basic tenets of the argument are discussed below.

Vitamin D synthesis is dependent on the transmission of UV into the lower layers of the epidermis. The annual number of sunshine hours decreases from the equator to the poles as does the intensity of UV. Moreover, in the far northerly regions, clouds, fogs and mists further obstruct the passage of UV rays. Thus, early hominids in their migrations to the northern latitudes would have been severely compromised in terms of vitamin D-synthesizing capacity. If they also had pigmented skins, it is highly probable that the already low quantities of UV would have been filtered out by the melanin before penetrating the deeper epidermis. The resultant lack of vitamin D formation would have led to rickets, a condition which would have brought about the extinction of early humans for two main reasons: first, the pelvic deformities of rickets would have made childbirth impossible and therefore curbed reproduction of the species; second, the distortion of the weight-bearing bones (bow legs and knock-knees) would seriously have hampered hunting ability (see Fig. 11.3). Therefore, according to the argument, human beings could only have survived the low UV availability of northern climes by evolving a depigmented skin to enable the scant available UV to traverse the stratum corneum (and beneath) and induce the conversion of 7-DHC into vitamin D. Hence arose the blond-haired, blue-eyed and fair-skinned people of Scandinavia and northern Europe.

An obvious weakness of the hypothesis is the case of the Inuit (Eskimo), who has a darker skin colour than the European Caucasoid and yet inhabits regions even further north which are sunless for much of the year. Murray (1934) explained that the Inuit diet was rich in fish and fish oils (foods with a high vitamin D value) and that this diet provided adequate vitamin D to forestall rickets. There was thus no need for skin whitening in the Inuit to compensate for UV impoverishment in the Arctic but, with this exception, the conventional human diet is not an important source of vitamin D and skin synthesis remains the major benefactor.

Loomis (1967) added a corollary to the hypothesis, namely, that the darkly pigmented skins of tropical people served to protect them against overproduction of vitamin D and resultant toxicity by reducing the epidermal transmission of UV-B. However, as mentioned above, vitamin D production in skin can never exceed certain limits irrespective of the intensity or duration of the ambient UV (Holick *et al.*, 1987). Long before experimental evidence became available it had been noted that vitamin D intoxication never occurred in either fair-skinned Caucasoids or albinos living in the tropics.

Evidence in support of the vitamin D hypothesis

Surveys conducted in the 1920s in the United States of America had noted the increased susceptibility of Negroids to rickets (Hess, 1930). It was found, for example, that among 2500 Caucasoids and 1500 Negroid women in labour, 13 per cent of the former and 40 per cent of the latter had deformity of the pelvis, a threefold difference. An examination of 500 Negroid and 500 Caucasoid children (ages 4 months to 4 years) in Memphis, Tennessee (Mitchell, 1930), revealed that rickets and severe rickets were twice and three times as common respectively in Negroid children. An autopsy study of children who died before the age of 2 years in Baltimore (during the years from 1926 to 1942) found little difference in the prevalence of rickets between Caucasoid and Negroid babies in the first year of life, but in the second year it was 45.5 per cent and 72.2 per cent respectively, with the more severe forms of the disease manifesting in Negroid babies (Follis, Park & Jackson, 1952).

The Negroid epidermis transmits three to five times less UV than the Caucasoid epidermis (see pp. 60–1), and it is therefore logical to assume that, as most of the previtamin D is formed by UV-B in the Malpighian layer of the epidermis, its synthesis would be reduced in black skin. Hess (1922) showed that if groups of black and white rats were given the minimal protective dose of light (the diet being the same in both groups), then the black rats and not the white ones developed rickets. Holick *et al.* (1981) exposed skin specimens to an amount of UV equivalent to that at the equator. Hypopigmented skin required 30 minutes to attain maximum previtamin D levels whereas deeply pigmented skin took up to 3.5

hours. In a subsequent experiment (Clemens *et al.*, 1982) they exposed Caucasoid and Negroid volunteers to one minimal erythemal dose of UV. The Caucasoids showed a marked rise in the blood concentration of vitamin D 1–2 days after irradiation but there was hardly any change in the Negroids. It was only after a Negroid subject received a six-fold increase in dose of UV that the vitamin D concentration increased to the same extent as in Caucasoids. The conclusion was that melanin restricted vitamin D synthesis by competing with 7-DHC for the UV light. These laboratory data are consistent with a number of recent North American surveys (Hollis & Pittard, 1984; Jacob *et al.*, 1984; Bell *et al.*, 1985; Specker, Tsang & Hollis, 1985a; Brazerol *et al.*, 1988) in which Negroid Americans had significantly lower blood and breast milk concentrations of vitamin D and 25-OHD than did Caucasoids. The findings were generally ascribed to the reduced skin synthesis of vitamin D in Negroids at these latitudes owing to their increased skin melanin. In a rural South African setting, however, the levels of 25-OHD in Negroid children were normal even though mild malnutrition was not uncommon in these children (Van der Westhuyzen, 1986).

The case against the vitamin D hypothesis
This hypothesis, as advanced by Murray (1934) and Loomis (1967), has attracted a great deal of attention and, although disputed by some, it has maintained its ascendancy as the most plausible explanation for the depigmentation of *Homo* north of 40° N. But, although melanin in the skin interferes with UV-B transmission to the sites of vitamin D synthesis, I maintain that this phenomenon is relatively unimportant with respect to the evolution of early European hominids.

In the first place, rickets is a disease associated with 'civilization'; it occurs in the smoke-polluted industrial cities of Europe (with their dark and overcrowded slums) and it is virtually absent in the rural areas. Sunlight is the most potent preventive agent against the disease and rickets is therefore almost non-existent among peoples living under natural conditions. A distinctive feature of early *Homo* is that he and she were exposed to the full impact of the natural environment. As their daily subsistence depended on gathering and hunting, they must have spent a considerable part of the day under the sun. It is difficult to date the advent of clothing (possibly 100 000 years ago) but even the wearing of animal skins and furs would have left extensive parts of the body exposed (e.g. arms and legs). And while the winters in Europe (particularly during the glacial periods) were exceptionally cold and dim, the late spring and summer would have afforded good opportunities for UV irradiation.

Several points are relevant in this discussion. One of the outstanding attributes of vitamin D is that it is capable of being stored in tissues such as fat and muscle (Rosenstreich, Rich & Volwiler, 1971; Mawer *et al.*, 1972) and retained there for use in times of shortage or privation. This storage phenomenon has been exploited therapeutically in that *single* large oral doses of vitamin D provide adequate cover for 6 months to 1 year in patients susceptible to vitamin D deficiency. In those who reside at latitudes above 50° N, so little (if any) vitamin D is formed in the skin during the winter months (Webb, Kline & Holick, 1988) that the winter blood 25-OHD concentrations are dependent on the values reached the previous summer. Caucasoid children living in the English Midlands had higher 25-OHD concentrations in winter if they had taken a seaside holiday the previous summer (Poskitt, Cole & Lawson, 1979). Gardeners in Dundee, Scotland (56.5° N), who spent their whole working day out of doors in both winter and summer showed peak 25-OHD levels in November and December (although their peak UV exposure had been in July): their December levels were nearly twice as high as those of indoor workers (Devgun *et al.*, 1981). People who therefore habitually receive UV light throughout the year – as early hominids would have done – continue to produce and store vitamin D during the autumn.

The studies demonstrating the peak production of vitamin D in summer and its storage for winter were performed on fair-skinned Caucasoids. The proponents of the vitamin D hypothesis would contend that in dark-skinned people, melanin would prevent the accrual of sufficient stores of the vitamin even during the northern summers and that there would be a deficiency in winter. This is the decisive argument if the hypothesis is to stand, but I believe that it fails.

The experiments cited above show unequivocally that Negroid skin is not so effective at synthesizing vitamin D as Caucasoid skin. But the data were relative rather than absolute. In other words, while white skin would produce its maximum quota of previtamin D after 30 minutes, black skin did so only after 3 hours; the administration of only one minimal erythemal dose of UV was required to increase the blood vitamin D level in Caucasoids, but it needed six times that dose to do the same in Negroids. It must be emphasized that the scenario in which early *Homo* evolved in Europe was one where the solar radiation in summer was present for prolonged periods and at reasonably high levels of intensity. With this degree of chronic exposure over several months and to a large body area, maximum quantities of vitamin D would have been formed and any resistance by the melanin barrier overcome.

There is a further factor that specifically applies to the higher latitudes and would certainly have been of importance during the Pleistocene Ice Ages – snow. Snow is a potent reflector of UV-B (up to 85 per cent), and it has been estimated for North America that between latitudes 40° N and 55° N the average continental snow cover on 31 March has a reflectivity (albedo) of 30–48 per cent (Kung, Pryson & Lenschow, 1964). At 45° N the intensity of UV-B at wavelengths 295–300 nm (the optimal range for photosynthesis of vitamin D) is about 35–45 times greater in April than it is in January but only about three to six times less than it is in the peak month of July (Johnson *et al.*, 1976). This means that in spring at 45° N the combination of increasing UV-B intensity and substantial snow reflectivity would constitute a fairly high field of UV radiation, possibly enough to induce sunburn in light-skinned individuals. There is little doubt that even darkly pigmented people would amass enough UV-B in these northern regions during the spring and summer to ensure adequate body stores of vitamin D for the remainder of the year.

If the above reasoning is correct, then it is contradicted by the findings mentioned earlier of (a) increased susceptibility to rickets and (b) lower blood levels of vitamin D and 25-OHD in Negroid Americans. This greater risk of vitamin D deficiency in Negroid Americans may relate more to non-biological factors than to the pigment blockade of UV light. Specker *et al.* (1985b) determined that only a small amount of sunlight exposure is necessary to maintain the blood 25-OHD concentrations of infants above the lower limit of the normal range: a conservative estimate was 30 minutes a week (wearing a diaper only) or 2 hours a week (fully clothed but without a cap). They found that, although breast-fed Negroid American babies under 6 months of age had significantly lower 25-OHD concentrations than breast-fed Caucasoid babies, the former had had negligible or zero UV exposure (Specker *et al.*, 1985b). The infants whose mothers were less educated were also those with lower sunlight exposures. Thus, the higher occurrence of rickets and vitamin D deficiency in Negroid Americans appears primarily to be due to socio-economic circumstances, namely, the occupancy of dark and overcrowded dwellings with the infants often being confined indoors and deprived of sunlight.

The vulnerability of British Asians to rickets and osteomalacia has been ascribed in part to their darker skin colour, but this idea is not upheld by observations that British residents of West Indian (Afro-Caribbean) origin, who have deeper skin pigmentation than the Asians, very rarely manifest clinical rickets (Cooke *et al.*, 1973; Holmes *et al.*, 1973; Goel *et al.*, 1976; Ford *et al.*, 1976). Moreover, artificial irradiation of Asian, Caucasoid and Negroid subjects with UV-B produced similar increases in blood 25-OHD levels irrespective of skin pigmentation (Stamp, 1975; Lo, Paris & Holick, 1986; Brazerol *et al.*, 1988). A study under natural conditions in Birmingham, England, revealed comparable increases in 25-OHD levels after the summer sunshine from March to October in groups of Asians, West Indians and Caucasoids (Ellis, Woodhead & Cooke, 1977). This absence of a blunted 25-OHD response to sunlight in the dark-skinned West Indians at high northerly latitudes (England lies farther north than the entire United States of America except for Alaska) proves that skin colour is not a major contributor to vitamin D deficiency in northern climes.

A very small amount of sun exposure appears necessary for the maintenance of normal vitamin D levels. At Tromsø, Norway (70° N, and well above the Arctic Circle), where the sun is below the horizon for two months of the year, subjects who did not take extra vitamin D supplementation showed a clear seasonal variation in 25-OHD concentrations but still had adequate levels by the end of winter (Vik, Try & Stromme, 1980). Coalminers in Newcastle, England, who worked an 8-hour underground day shift from May to July did not differ in blood 25-OHD levels from surface workers and non-miners (Shuster *et al.*, 1981).

There are two geographical areas in the world of particular relevance to this discussion. One is the Dead Sea in Israel (400 metres below sea level and the lowest point on earth), which even at midday in the height of summer receives very weak UV-B (Kushelevsky & Slifkin, 1975). The other is the equatorial rain forest belt of Central Africa, home of the Bambuti Pygmies who live for much of the time under the thick forest canopy where very little UV light can penetrate. There is no scientific evidence to suggest that rickets or osteomalacia has featured in either of these places.[6] The short stature of the Pygmies appears to be an adaptation to the equatorial forest biome (Hiernaux, Rudan & Brambati, 1975); it is not due to vitamin D deficiency but probably to a defect in the growth hormone system such as that postulated by Merimee, Zapf & Froesch (1981).

It is also useful to review the archaeological record and to judge the frequency of rickets and osteomalacia in prehistoric populations. In 1872 Virchow thought that the Neanderthal skull showed signs of rickets, and this observation triggered the proposal that the Neanderthals of Europe suffered from that disorder (Ivanhoe, 1970). These Neanderthals lived through the first part of the Würm Ice Age in Europe – from 70 000 BP until 35 000 BP when they became extinct – under climatic circumstances that would have favoured the development of rickets. The atmospheric turbulence, greater cloud cover and increased precipitation would have attenuated the UV reaching the earth, and the extremely cold conditions would have driven the Neanderthals into caves and made them wear thick furs. The absence of fatty fish from their diet (fishing technology had not been developed) would have deprived them of a dietary source of vitamin D. Indeed, Ivanhoe (1970) claimed that every Neanderthal child skull he had examined showed the stigmata of severe rickets (but his paper is poorly documented and not generally accepted). The consensus is that Neanderthal characteristics are not due to rickets (Mayr & Campbell, 1971) and that the typical European Neanderthal skeleton was so widely distributed in both time (including during the warm Riss–Würm interglacial) and place (including the Middle East) that vitamin D deficiency could not have been responsible.

Rickets has been described sporadically in European skeletal material (predominantly from Scandinavia and Hungary) dating from the Neolithic period until medieval times (Wells, 1975). However, the prevalence of rickets appears to have been very low; it occurred in no more than 1 per cent of skeletons from Swedish and Danish medieval cemeteries (AD 1100–1550) and there was no trace of it in Anglo-Saxon remains from East Anglia. The disease gradually increased during the post-medieval era of urbanization and rapidly escalated with the advent of industrialization, so that by the eighteenth and nineteenth centuries it was rampant in skeletal surveys. This again highlights the fact that rickets is not a 'natural' disorder but arises from abnormal environmental conditions. In the medieval cities of Hungary it was three times more common in skeletons from graveyards of the twelfth century than from those of the ninth, no doubt on account of the overcrowded, windowless houses that had mushroomed during that period (Zivanovic, 1982). The discovery of rickets in certain Cro-Magnon skeletons from Palaeolithic and Mesolithic settlements in the Danube valley was subsequently explained on the basis that these settlements were at the bottom of a gorge which was overshadowed by dense forest and therefore deprived of sunlight (Zivanovic, 1982).

Outside Europe early skeletal evidence of rickets is very sparse. In pre-Columbian North and South America it was, except for occasional and questionable reports, virtually non-existent (Wells, 1975). In a survey of 457 skeletons from an Iroquoian ossuary (dated AD 1490) 65 miles northeast of Toronto, no mention was made of the presence of rickets or osteomalacia in any of the specimens (Pfeiffer, 1984). The absence of overt vitamin D deficiency in prehistoric North American peoples is of interest because they first entered the continent during the Ice Age (at least 11 500 years ago and, according to some estimates, 20 000–40 000 years ago) when they were exposed to a similar set of climatic and ecological forces as the Neanderthals. Moreover, as judged by present-day Amerindian populations, their skin colour would have been darker than that of European Caucasoids (see Table 7.2). The occurrence of moderately pigmented and apparently rickets-free *Homo* within the northern zones of America during the Ice Age is testimony against the vitamin D hypothesis of depigmentation. Another pertinent palaeo-pathological survey examined 92 skeletons from an Early Bronze Age site (4000–5000 BP) near the shores of the Dead Sea (Ortner, 1979). There is no reference to rickets or osteomalacia in spite of the very low UV-B levels at the Dead Sea. On the other hand, signs of rickets were not uncommon in a skeletal analysis of a free black community in Phila-delphia (1823–1841) which existed under a variety of stresses, including poor living conditions and limited exposure to the sun (Angel *et al.*, 1987).

The upshot of all the arguments raised here is that the hypothesis of Murray (1934) and Loomis (1967) does not survive critical scrutiny. Even assuming from the transmission characteristics that black epidermis would always cause a shortfall in vitamin D formation compared to white epidermis, Beadle (1977) calculated that Negroids would still produce a total annual amount of vitamin D that fell within the normal recom-mended range right up to the latitudes of the Arctic Circle – and this estimate was based on an exposure of only 10.5 per cent of the body area (head, neck and hands). Furthermore, there was no synthesis of vitamin D in Caucasoid skin specimens exposed to sunlight in Boston (42.2° N) from November to February (inclusive) and in Edmonton (52° N) from October to March (inclusive) (Webb *et al.*, 1988). However, during the same period of exposure in Boston (November to February), there was a significant photodegradation within the specimens of the existing vitamin D stores (Webb, De Costa & Holick, 1989).[6] The major thrust of the hypothesis is that skin depigmentation evolved in the north to utilize the winter sunlight for vitamin D synthesis. As 'white' skin during the northern winter appears not only incapable of such synthesis but also likely to promote *depletion* of vitamin D, the hypothesis is severely undermined.

In summary, rickets is a disease of industrialization, urbanization and overpopulation; it is a disease which stems from a deprivation of sun-shine. The least likely backdrop from which rickets would emerge was the open-air, hunter–gatherer lifestyle of naked or semi-naked *Homo* on the Eurasian and North American landscapes during the Pleistocene. Under

such circumstances there is not a scintilla of palaeontological or experimental evidence that rickets (or osteomalacia) ever did or would have
manifested, regardless of skin colour. Pigmented skin may not photosyn-
thesize vitamin D as readily as white skin under certain conditions but,
given the prolonged hours of sun exposure in the northern summer and
the body's remarkable potential for storing the compound, early pigmented humans would have manufactured and stored the vitamin as
efficiently as their depigmented counterparts.

There is also a dietary aspect to the scenario. After the retreat of the Ice
Age in Europe about 10 000 years ago, *Homo* moved into more northerly
parts (e.g. Scandinavia) which had previously then been ice-bound and
uninhabitable. The species had by then crafted a basic fishing technology
which it continued to develop; and postglacial Europeans shifted their
activities from foraging land resources to harvesting marine life (Zvelebil, 1986). Fatty fish such as salmon and eels (a source rich in vitamin D)
formed a significant part of the diet. Ironically, the salmon cult became so
paramount a feature of life in some north-western coastal regions of
Canada that the anticipated hazard was not rickets but vitamin D
intoxication (Lazenby & McCormack, 1985)! Recent isotope studies
have disclosed that 6000–9000 years ago the Danes subsisted on a diet
dominated by sea food, although this pattern subsequently changed
(Tauber, 1981). Moreover, during certain Upper Palaeolithic times in
Europe salmon were an abundantly available food supply, from, for
example, the Dordogne river in France. Thus, this dietary contribution
would have diminished the need for endogenous (skin) synthesis of
vitamin D and its presence further weakens some of the evolutionary
arguments for depigmentation in the northernmost regions.

Cold injury

Perhaps a more likely hypothesis for the evolution of a depigmented skin
in the north is set out in a paper by Post, Daniels & Binford (1975). It is
structured on the observation that Negroids are apparently more susceptible to cold injury than Caucasoids. Their paper cites numerous references to frostbite or trench foot in Negroid troops during the various wars
of the twentieth century. For example, there were 1225 cases of 'frozen
feet' among Senegalese troops in France during only two days in mid-
April 1917 – whereas at the same time cold injury was negligible among
the French troops. Approximately one out of every three cases of trench
foot suffered by the United States Army during the 1944–45 Italian
campaign came from a Negroid infantry division. During the Korean War
the casualty rate from frostbite was three to six times greater among

Negroid soldiers than among Caucasoids and, although the injuries followed the same distribution in Negroids and Caucasoids, they were more severe in the former.

The hypothesis goes on to postulate a direct relationship between degree of cold susceptibility and amount of skin pigmentation. Animal experiments show that melanocytes are more easily destroyed by freezing techniques than the skin itself and, even in the human situation, cold injury and cryosurgery induce skin depigmentation. Post *et al.* (1975) used piebald guinea-pigs to test the effects of freezing on the pigmented and non-pigmented skin of the same animals. Microscopy of the frozen specimens revealed that, although damage was evident at both the pigmented and non-pigmented sites, it was more severe in the pigmented skin. In addition, areas of lightly pigmented skin showed a degree of injury that was intermediate between that of dark and of non-pigmented areas. The authors have also reported unpublished findings from other workers that pigmented melanoma cells *in vitro* have a much lower survival rate after freezing than do non-pigmented cells.

This frostbite hypothesis introduces another dimension into the possible selection pressures in the north which favoured reduced skin pigmentation. Temperature is the variable, second only in importance to UV, that contributes to variations in skin pigmentation (Roberts & Kahlon, 1976). Frostbite causes crippling of the hands and feet, and these disabilities could seriously endanger the survival of hunter–gatherers of reproductive age (Steegmann, 1967). Moreover, death may ensue from secondary infection such as gangrene or tetanus. As with sunburn, frostbite in the infant would be a decisive selective force – particularly if it involved the face, because frozen cheeks and lips would prevent sucking at the breast with resultant dehydration and death. A negative effect on reproductive fitness would be frostbite of the penis, and there is an amusing self-report of a mild variant of this condition (Hershkowitz, 1977) after a half-hour jog one North American winter evening during which the air temperature of $-8°$ C was combined with severe wind-chill factor. It is relevant that much of early *Homo*'s evolution in Europe occurred during the Ice Ages when the ambient temperatures were about 8–12 degrees colder than at present. These extraordinarily cold conditions would have swiftly selected against pigmented skins if the frostbite hypothesis is correct. Even in southern Africa temperatures dropped during the last Ice Age and, at its peak (18 000 BP), were 5–9 degrees lower than they are today. It is just possible that the Khoisan peoples (formerly Hottentot and Bushman), who roamed through southern Africa within the last glacial period, are yellow-brown in pigmentation (and much lighter than African Negroid populations) because this skin colour was more compatible with the cold conditions obtaining in those regions.

One major problem with the frostbite hypothesis is that although experimental studies have confirmed that Negroids show poorer physiological tolerance to cold testing than Caucasoids, the latter have inferior responses to the Inuit and Amerindians (Steegmann, 1975), who are darker in skin pigmentation than Caucasoids. This observation tends to conflict with the argument that the presence of melanin is the factor increasing the vulnerability to cold injury. It is possible that the Inuit and Amerindians have a superimposed genetic superiority in cold tolerance which derives from their prehistoric adaptation to the bitterly cold climate of continental Asia (a much severer one than the maritime Western European counterpart) (Steegmann, 1975). Caution is advised before attributing variations in cold tolerance to ethnic-related differences. The amount of subcutaneous fat, for example, may be more important than inherited physiological differences in conditioning the responses to cold stress (Hanna & Smith, 1975).

Conclusions

This has been a long and difficult chapter, one that has been based more on speculation than on fact. As research generates new findings these will pose more questions and evoke different hypotheses. The evolution of skin colour will be an ongoing conundrum for a long time.

Although the main theories of skin colour evolution have been reviewed, there are several other perspectives that could have been debated. One of these highlights the role of folic acid, an essential vitamin that is destroyed by UV light. The suggestion is that in areas of high solar radiation dark skins evolved to protect against light-induced folate deficiency – a serious problem which might affect reproductive capacity (Branda & Eaton, 1978) – but the hypothesis is insufficiently structured and the data provided are unconvincing. There is no evidence, for instance, that Caucasoids or albinos in tropical regions are more folate-deficient than Negroids; on the contrary, frank folate deficiency is so prevalent – for example, among South African Negroids (Van der Westhuyzen *et al.*, 1986; Baynes *et al.*, 1986) – that it is difficult to credit a dark skin with any protective function against such deficiency.

Then there is the hypothesis of Guthrie (1970) that dark coloration in the human is a social display of threat: females are more lightly coloured than males because non-aggression was an attribute selected for in females. Depigmentation evolved in more integrated and interdependent societies where it was necessary to bring about 'the general deemphasis of the ancestral threat devices'. It should be noted that Hamilton (1973) has challenged this viewpoint because it assumes that black people were in a relatively poor state of social organization during the evolution of skin colour, an assumption that is at variance with the high levels of social cohesiveness known to have existed among indigenous tropical inhabitants.

Skin colour is a polygenic trait for which there may be several selective advantages and disadvantages. The question is often raised as to how much of evolutionary time it would take for significant differences in the skin colour of populations to emerge. Livingstone (1969) constructed a computer model on the assumption that four genes were involved in the determination of skin colour. He estimated that in 800 generations without gene dominance, and 1500 generations with gene dominance, the whole spectrum of skin colour differences from black to white could have evolved (i.e. in a period of between 20 000 and 40 000 years). The evolution of an intermediate skin colour (brownish) from either white or black ancestors would have required about 500 generations. The speed of evolutionary change can be much faster, as demonstrated by the house sparrow which, since its introduction into North America from England and Germany in 1852, has undergone distinct geographical variations in colour in not more than 111 generations (Johnston & Selander, 1964). It is difficult to make estimates of time spans, as these depend on numerous factors, but significant changes in certain human physical characteristics have been accomplished in a matter of 30 generations (750 years).

It has been recognized for centuries that skin colour and sunlight are associated, and this is exactly the current position. The one statement that can be made with reasonable confidence is that black skins are adapted to high levels, and white skins to low levels, of UV light. The specific reasons for this association have yet to be resolved and there is an urgent need for better-orientated data and research. An area that may yield some exciting clues is that of photoimmunology (see p. 45). UV suppresses immunity, and it is theoretically possible that the presence or absence of skin melanin may regulate or modulate certain immune processes. Perhaps this phenomenon holds the key to our understanding of the most visible, the most abused, and one of the most enigmatic features of the human animal.

Notes

1 There is a growing consensus that Pleistocene hominids may have obtained meat not primarily by game hunting but through opportunistic scavenging of

the carcasses of large mammals abandoned by other predators (Speth, J. D. (1989). *Journal of Human Evolution*, **18**, 329–43).

2 A counter-argument to this might be that when the poor hunters stayed at base to chip stones for tools, they would have had ample opportunities to copulate while the good hunters were away. A similar proposal has been made with regard to male Hopi albinos (see note 3, Chapter 10).

3 A good historical example of 'anti-camouflage' in warfare is that of the British 'Redcoats' in South Africa, whose conspicuousness in scarlet uniforms against the bushveld landscape is said to have cost them many a battle!

4 Scientific racism was blatant in its attempts to explain the extraordinarily low mortality experienced by Negroids in the antebellum American South to malaria and yellow fever, diseases which annihilated Caucasoids. Negroids were deemed to belong to a different species – and an inferior one at that (Kiple & King, 1981)!

5 Not only does scant UV reach the forest floor but the Pygmies are also dark-skinned, although less so than the surrounding Negroid populations. One theory for the lack of rickets in Bambuti Pygmies is that their diet consists of an abundance of insects (e.g. termites and caterpillars), the fat of which may supply their vitamin D requirement (Fischer, E. (1961). *Zeitschrift für Morphologie und Anthropologie*, **51**, 119–36).

6 The explanation given (Webb *et al.*, 1989) is that, during the Boston winter, the shorter UV-B radiation (<315 nm) – which is chiefly responsible for the photosynthesis of vitamin D from 7-DHC – is virtually screened out before it reaches the earth's surface, whereas the longer UV wavelengths (up to 330 nm) are present in sufficient amounts to produce photodestruction of skin vitamin D.

References

Akutsu, Y. & Jimbow, K. (1988). Immunoelectron microscopic demonstration of human melanosome-associated antigens (HMSA) on melanoma cells: comparison with tyrosinase distribution. *Journal of Investigative Dermatology,* **90**, 179–84.

Allen, B. M. (1916). The results of extirpation of the anterior lobe of the hypophysis and of the thyroid of *Rana pipiens* larvae. *Science,* **44**, 755–8.

Ananth, J. & Yassa, R. (1982). Tardive dyskinesia and skin pigmentation. *British Journal of Psychiatry,* **141**, 194–5.

Angel, J. L., Kelley, J. O., Parrington, M. & Pinter, S. (1987). Life stresses of the free black community as represented by the first African Baptist Church, Philadelphia, 1823–1841. *American Journal of Physical Anthropology,* **74**, 213–29.

Anonymous (1973). Op to turn 'black' girl white again. *Star* (Johannesburg, November 13).

Apkarian, P., Reits, D., Spekreijse, H. & Van Dorp, D. (1983). A decisive electrophysiological test for human albinism. *Electroencephalography and Clinical Neurophysiology,* **55**, 513–31.

Ardener, E. W. (1954). Some Ibo attitudes to skin pigmentation. *Man,* **54**, 71–3.

Arendt, J., Aldhous, M. & Marks, V. (1986). Alleviation of jet lag by melatonin: preliminary results of controlled double blind trial. *British Medical Journal,* **292**, 1170.

Ayala Uribe, M. G. (1976). The Mongoloid spot. *Australian Journal of Dermatology,* **17**, 61–4.

Bagnara, J. T., Matsumoto, J., Ferris, W. *et al.* (1979). Common origin of pigment cells. *Science,* **203**, 410–15.

Baker, P. T. (1958). Racial differences in heat tolerance. *American Journal of Physical Anthropology,* **16**, 287–305.

Baldwin, J. C. & Damon, A. (1973). Some genetic traits in Solomon Island populations. V. Assortative mating, with special reference to skin colour. *American Journal of Physical Anthropology,* **39**, 195–202.

Ball, W. A. & Caroff, N. C. (1986). Retinopathy, tardive dyskinesia, and low-dose thioridazine. *American Journal of Psychiatry,* **143**, 256–7.

Banerjee, S. (1984). The inheritance of constitutive and facultative skin colour. *Clinical Genetics,* **35**, 256–8.

(1985). Assortative mating for colour in Indian populations. *Journal of Biosocial Science,* **17**, 205–9.

Barber, J. I., Townsend, D., Olds, D. P. & King, R. A. (1984). Dopachrome oxidoreductase: a new enzyme in the pigment pathway. *Journal of Investigative Dermatology,* **83**, 145–9.

Barden, H. (1969). The histochemical relationship of neuromelanin and lipofuscin. *Journal of Neuropathology and Experimental Neurology,* **28**, 419–41.

Barnes, R. B. (1963). Thermography of the human body. *Science,* **140**, 870–7.

Barnicot, N. A. (1958). Reflectometry of the skin in southern Nigerians and in some mulattoes. *Human Biology,* **30**, 150–60.

Barr, F. E. (1983). Melanin: the organizing molecule. *Medical Hypotheses,* **11**, 1–140.

Barr, R. D., Rees, P. H., Cordy, P. E. *et al.* (1972). Nephrotic syndrome in adult Africans in Nairobi. *British Medical Journal,* **2**, 131–4.

Barrenäs, M-L. & Lindgren, F. (1990). The influence of inner ear melanin on the susceptibility to TTS in humans. *Scandinavian Audiology,* **19**, 97–102.

Barron, J. A. (1986). Cigarette smoking and Parkinson's disease. *Neurology,* **36**, 1490–6.

Barton, D. E., Kwon, B. S. & Francke, U. (1988). Human tyrosinase gene, mapped to chromosome 11 (q14–q21), defines second region of homology with mouse chromosome 7. *Genomics,* **3**, 17–24.

Bastide, R. (1968). Color, racism and Christianity. In *Color and Race,* ed. J. H. Franklin, pp. 34–49. Boston: Beacon Press.

Baynes, R. D., Meriwether, W. D., Bothwell, T. H. *et al.* (1986). Iron and folate status of pregnant black women in Gazankulu. *South African Medical Journal,* **70**, 148–51.

Beadle, P. C. (1977). The epidermal biosynthesis of cholecalciferol (vitamin D_3). *Photochemistry and Photobiology,* **25**, 519–27.

Becker, S. W. (1959). Historical background of research on pigmentary diseases of the skin. *Journal of Investigative Dermatology,* **32**, 185–96.

Beckham, A. S. (1946). Albinism in Negro children. *Journal of Genetic Psychology,* **69**, 199–215.

Bell, N. H., Greene, A., Epstein, S. *et al.* (1985). Evidence for alteration of the vitamin D-endocrine system in blacks. *Journal of Clinical Investigation,* **76**, 470–3.

Benedetto, J-P., Ortonne, J-P., Voulot, C. *et al.* (1981, 1982). Role of thiol compounds in mammalian melanin pigmentation: parts I and II. *Journal of Investigative Dermatology,* **77**, 402–5; **79**, 422–4.

Benjannet, S., Seidah, N. G., Routhier, R. & Chretien, M. (1980). A novel human pituitary peptide containing the γ-MSH sequence. *Nature,* **285**, 415–16.

Benning, T. L., McCormack, K. M., Ingram, P., Kaplan, D. L. & Shelburne, M. D. (1988). Microprobe analysis of chlorpromazine pigmentation. *Archives of Dermatology,* **124**, 1541–4.

Bental, E. (1979). Observations on the electroencephalogram and photosensitivity of South African black albinos. *Epilepsia,* **20**, 593–7.

Best, D. L., Field, J. T. & Williams, J. E. (1976). Color bias in a sample of young German children. *Psychological Reports,* **38**, 1145–6.

Best, D. L., Naylor, C. E. & Williams, J. E. (1975). Extension of color bias research to young French and Italian children. *Journal of Cross-Cultural Psychology,* **6**, 390–405.

Billingham, R. E. (1948). Dendritic cells. *Journal of Anatomy,* **82**, 93–109.

Binder, R. L., Kazamatsuri, H., Nishimura, T. & McNiel, D. E. (1987). Tardive dyskinesia and neuroleptic-induced Parkinsonism in Japan. *American Journal of Psychiatry,* **144**, 1494–6.

Binder, R. L. & Levy, R. (1981). Extrapyramidal reactions in Asians. *American Journal of Psychiatry*, **138**, 1243–4.

Birbeck, M. S. & Barnicot, N. A. (1959). Electron microscope studies on pigment formation. In *Pigment Cell Biology*, ed. M. Gordon, pp. 549–57. New York: Academic Press.

Birdsell, J. B. (1950). Some implications of the genetical concept of race in terms of spatial analysis. *Cold Spring Harbor Symposium on Quantitative Biology*, **15**, 259–314.

Bischitz, P. G. & Snell, R. S. (1960). A study of the effect of ovariectomy, oestrogen and progesterone on the melanocytes and melanin in the skin of the female guinea-pig. *Journal of Endocrinology*, **20**, 312–19.

Bjerring, P. & Andersen, P. H. (1987). Skin reflectance spectrophotometry. *Photodermatology*, **4**, 167–71.

Black, G., Matzinger, E. & Gange, R. W. (1985). Lack of photo-protection against UVB-induced erythema by immediate pigmentation induced by 382 nm radiation. *Journal of Investigative Dermatology*, **85**, 448–9.

Blum, H. F. (1961). Does the melanin pigment of human skin have adaptive value? *Quarterly Review of Biology*, **36**, 50–63.

Bocchetta, A. & Corsini, G. U. (1986). Parkinson's disease and pesticides. *Lancet*, **2**, 1163.

Bogerts, B. (1981). A brainstem atlas of catecholamine neurons in man, using melanin as a natural marker. *Journal of Comparative Neurology*, **197**, 63–80.

Bolt, A. G. & Forrest, I. S. (1968). *In vivo* and *in vitro* interaction of chlorpromazine and melanin. *Recent Advances in Biological Psychiatry*, **10**, 20–8.

Bortz, W. M. (1985). Physical exercise as an evolutionary force. *Journal of Human Evolution*, **14**, 145–55.

Boswell, D. A. & Williams, J. E. (1975). Correlates of race and color bias among preschool children. *Psychological Reports*, **36**, 147–54.

Boulton, A. A., Yu, P. H. & Tipton, K. F. (1988). Biogenic amine adducts, monoamine oxidase inhibitors, and smoking. *Lancet*, **1**, 114–15.

Boyle, J., Mackie, R. M., Briggs, J. D., Junor, B. J. & Aitchison, T. C. (1984). Cancer, warts, and sunshine in renal transplant patients. *Lancet*, **1**, 702–4.

Brain, C. K. (1965). Observations on the behaviour of vervet monkeys, *Cercopithecus aethiops*. *Zoologica Africana*, **1**, 13–27.

Branda, R. F. & Eaton, J. W. (1978). Skin color and nutrient photolysis: an evolutionary hypothesis. *Science*, **201**, 625–6.

Bräuer, G. & Chopra, V. P. (1980). Estimating the heritability of hair colour and eye colour. *Journal of Human Evolution*, **9**, 625–30.

Brazerol, W. F., McPhee, A. J., Mimouni, F., Specker, B. L. & Tsang, R. C. (1988). Serial ultraviolet B exposure and serum 25 hydroxyvitamin D response in young adult American blacks and whites: no racial differences. *Journal of the American College of Nutrition*, **7**, 111–18.

British Medical Journal (1971). Adrenaline into melanin. Editorial. *British Medical Journal*, **2**, 486.

Brown, C. (1984). *Black and White Britain: The Third PSI Survey*. London: Heinemann.

Brown, J. D. & Doe, R. P. (1978). Pituitary pigmentary hormones. *Journal of the American Medical Association*, **240**, 1273–8.

Büchi, E. C. (1957). Eine spektrophotometrische Untersuchung der Hautfarbe von Angehörigen verschiedener Kasten in Bengalen. *Bulletin der Schweizerische Gesellschaft für Anthropologie und Ethnologie,* **34**, 7–8.

Buckler, H. M., Freeman, H., Chetty, M. C. & Anderson, D. C. (1988). Nelson's syndrome and behavioural changes reversed by selective adenomectomy. *British Journal of Psychiatry,* **152**, 412–14.

Buckley, W. R. & Grum, F. (1961). Reflection spectrophotometry: use in evaluation of skin pigmentary disturbances. *Archives of Dermatology,* **83**, 249–61.

Burtt, E. H. & Gatz, A. J. (1982). Color convergence: is it only mimetic? *American Naturalist,* **119**, 738–40.

Byard, P. (1981). Quantitative genetics of human skin color. *Yearbook of Physical Anthropology,* **24**, 123–37.

Byard, P. J. & Lees, F. C. (1981). Estimating the number of loci determining skin colour in a hybrid population. *Annals of Human Biology,* **8**, 49–58.

Byard, P. J. & Lees, F. C. (1982). Skin colorimetry in Belize. II. Inter- and intra-population variation. *American Journal of Physical Anthropology,* **58**, 215–19.

Bystryn, J-C. & Pfeffer, S. (1988). Vitiligo and antibodies to melanocytes. In *Advances in Pigment Cell Research,* ed. J. T. Bagnara, pp. 192–206. New York: Alan R. Liss.

Caro, L. (1980). La reflectancia de la piel en Españoles del Noroeste. *Boletín de la Sociedad Española de Antropologia Biologica,* **1**, 24–31.

Carruthers, R. (1966). Chloasma and oral contraceptives. *Medical Journal of Australia,* **2**, 17–20.

Carstam, R., Hansson, C., Krook, G. *et al.* (1985). Oxidation of dopa in human albinism. *Acta Dermatovenereologica,* **65**, 435–7.

Cartwright, R. A. (1975). Skin reflectance results from Holy Island, Northumberland. *Annals of Human Biology,* **2**, 347–54.

Castle, D., Kromberg, J., Kowalsky, R. *et al.* (1988). Visual evoked potentials in Negro carriers of the gene for tyrosinase-positive oculocutaneous albinism. *Journal of Medical Genetics,* **25**, 835–7.

Chamla, M-C. & Demoulin, F. (1978). Réflectance de la peau, pigmentation des cheveux et des yeux des Chaouïas de Bouzina (Aurès, Algérie). *L'Anthropologie* (Paris), **82**, 61–94.

Chapman, L. J., Peters, H. A., Matthews, C. G. & Levine, R. L. (1987). Parkinsonism and industrial chemicals. *Lancet,* **1**, 332–3.

Cherubino, M. (1972). Light, camouflage, structure and control of the melanocytes in the inner ear. *Research Progress in Organic–Biological and Medicinal Chemistry,* **3**, 754–63.

Clark, P., Stark, A. E., Walsh, R. J., Jardine, R. & Martin, N. G. (1981). A twin study of skin reflectance. *Annals of Human Biology,* **8**, 529–41.

Clarke, D. J. & Buckley, M. E. (1989). Familial association of albinism and schizophrenia. *British Journal of Psychiatry,* **155**, 551–3.

Clemens, T. L., Adams, J. S., Henderson, S. L. & Holick, M. F. (1982). Increased skin pigment reduces the capacity of skin to synthesize vitamin D_3. *Lancet,* **1**, 74–6.

Clive, D. & Snell, R. S. (1969). Effect of melatonin on mammalian hair colour. *Journal of Investigative Dermatology*, **53**, 159–62.

Cole, G. F., Conn, P., Jones, R. B., Wallace, J. & Moore, V. R. (1987). Cognitive functioning in albino children. *Developmental Medicine and Child Neurology*, **29**, 659–65.

Collins, R. N., Lerner, A. B. & McGuire, J. S. (1966). The relationship of skin color to zygosity in twins. *Journal of Investigative Dermatology*, **47**, 78–82.

Conlee, J. W., Abdul-Baqi, K. J., McCandless, G. A. & Creel, D. J. (1986a). Differential susceptibility to noise-induced permanent threshold shift between albino and pigmented guinea pigs. *Hearing Research*, **23**, 81–91.

Conlee, J. W., Parks, T. N. & Creel, D. J. (1986b). Reduced neuronal size and dendritic length in the medial superior olivary nucleus of albino rabbits. *Brain Research*, **363**, 28–37.

Conway, D. L. & Baker, P. T. (1972). Skin reflectance of Quecha Indians: the effects of genetic admixture, sex and age. *American Journal of Physical Anthropology*, **36**, 267–82.

Cook, L. M., Mani, G. S. & Varley, M. E. (1986). Postindustrial melanism in the peppered moth. *Science*, **231**, 611–13.

Cooke, W. T., Swan, C. H., Asquith, P., Melikan, V. & McFeely, W. E. (1973). Serum alkaline phosphatase and rickets in urban schoolchildren. *British Medical Journal*, **1**, 324–7.

Cope, Z. (1964). Jane Austen's last illness. *British Medical Journal*, **2**, 182–3.

Corsellis, J. A. (1953). The pathological report of a case of phenylpyruvic oligophrenia. *Journal of Neurology, Neurosurgery and Psychiatry*, **16**, 139–43.

Cosnett, J. E. & Bill, P. L. (1988). Parkinson's disease in blacks: observations on epidemiology in Natal. *South African Medical Journal*, **73**, 281–3.

Costas, R., Garcia-Palmieri, M. R., Sorlie, P. & Hertzmark, E. (1981). Coronary heart disease risk factors in men with light and dark skin in Puerto Rico. *American Journal of Public Health*, **71**, 614–19.

Cotzias, G. C., van Woert, M. H. & Schiffer, L. M. (1967). Aromatic amino acids and modification of Parkinsonism. *New England Journal of Medicine*, **276**, 374–9.

Cowie, V. & Penrose, L. S. (1951). Dilution of hair colour in phenylketonuria. *Annals of Eugenics*, **150**, 297–301.

Cowles, R. B. (1959). Some ecological factors bearing on the origin and evolution of pigment in the human skin. *American Naturalist*, **93**, 283–93.

Creel, D. (1980). Inappropriate use of albino animals as models in research. *Pharmacology, Biochemistry and Behavior*, **12**, 969–77.

Creel, D., Garber, S. R., King, R. A. & Witkop, C. J. (1980). Auditory brainstem anomalies in human albinos. *Science*, **209**, 1253–5.

Crome, L. (1962). The association of phenylketonuria with leucodystrophy. *Journal of Neurology, Neurosurgery and Psychiatry*, **25**, 149–53.

D'Amato, R. J., Alexander, G. M., Schwartzman, R. J. *et al.* (1987). Evidence for neuromelanin involvement in MPTP-induced neurotoxicity. *Nature*, **327**, 324–6.

Dahlberg, F. (1981). *Woman the Gatherer*. New Haven: Yale University Press.

Daniels, F., Post, P. W. & Johnson, B. E. (1972). Theories of the role of pigment in the evolution of human races. In *Pigmentation: its Genesis and Control,* ed. V. Riley, pp. 13–22. New York: Appleton-Century-Crofts.

Das, S. R. & Mukherjee, D. P. (1963). A spectrophotometric colour survey among four Indian castes and tribes. *Zeitschrift für Morphologie und Anthropologie,* **54**, 190–200.

Davey, A. G. & Norburn, M. V. (1980). Ethnic awareness and ethnic differentiation amongst primary school children. *New Community,* **8**, 51–60.

De Long, S. L. (1968). Incidence and significance of chlorpromazine-induced eye changes. *Diseases of the Nervous System,* **29**, supplement, 19–22.

Dean, G. (1963). *The Porphyrias: Story of Inheritance and Environment.* London: Pitman Medical.

Derbes, V. J., Fleming, G. & Becker, S. W. (1955). Generalized cutaneous pigmentation of diencephalic origin. *Archives of Dermatology,* **72**, 13–22.

Devgun, M. S., Paterson, C. R., Johnson, B. E. & Cohen, C. (1981). Vitamin D nutrition in relation to season and occupation. *American Journal of Clinical Nutrition,* **34**, 1501–4.

Diffey, B. L., Larko, O. & Swanbeck, G. (1982). UV-B doses received during different outdoor activities and UV-B treatment of psoriasis. *British Journal of Dermatology,* **106**, 33–41.

Dmiel, R., Prevulotzky, A. & Shkolnik, A. (1980). Is a black coat in the desert a means of saving metabolic energy? *Nature,* **283**, 761–2.

Dodt, E., Copenhaver, R. M. & Gunkel, R. D. (1959). Electroretinographic measurement of the spectral sensitivity in albinos, Caucasians and Negroes. *Archives of Ophthalmology,* **63**, 795–803.

Dogliotti, M., Caro, I., Hartdegen, R. G. & Whiting, D. A. (1974). Leucomelanoderma in blacks: a recent epidemic. *South African Medical Journal,* **48**, 1555–8.

Dornhurst, A., Jenkins, J. S., Lamberts, S. W. *et al.* (1983). The evaluation of sodium valproate in the treatment of Nelson's syndrome. *Journal of Clinical Endocrinology and Metabolism,* **56**, 985–91.

Ducros, A., Ducros, J. & Robbe, P. (1975). Pigmentation et brunissement comparés d'Eskimo (Ammassalimiut, Groenland De L'Est). *L'Anthropologie* (Paris), **79**, 299–316.

Dupré, A., Ortonne, J-P., Viraben, R. & Arfeux, F. (1985). Chloroquine-induced hypopigmentation of hair and freckles. *Archives of Dermatology,* **121**, 1164–6.

Eady, R. A., Gunner, D. B., Garner, A. & Rodeck, C. H. (1983). Prenatal diagnosis of oculocutaneous albinism by electron microscopy of fetal skin. *Journal of Investigative Dermatology,* **80**, 210–12.

Eberle, A. N. (1988). *The Melanotropins: Chemistry, Physiology and Mechanisms of Action.* Basel: Karger.

Edwards, E. A. & Duntley, S. Q. (1939a). An analysis of skin pigment changes after exposure to sunlight. *Science,* **90**, 235–7.

(1939b). The pigments and color of living human skin. *American Journal of Anatomy,* **65**, 1–33.

(1949). Cutaneous vascular changes in women in reference to the menstrual cycle and ovariectomy. *American Journal of Obstetrics and Gynaecology,* **57**, 501–9.

Ellis, G., Woodhead, J. S. & Cooke, W. T. (1977). Serum-25-hydroxyvitamin-D concentrations in adolescent boys. *Lancet,* **1**, 825–8.

Emiru, V. P. (1971). Response to mydriatics in the African. *British Journal of Ophthalmology,* **55**, 538–43.

Ephraim, A. I. (1959). On sudden or rapid whitening of the hair. *Archives of Dermatology,* **79**, 228–35.

Everett, M. A., Nordquist, R. & Wasik, R. (1979). Melanosome size and distribution in American Indians. In *Pigment Cell,* Vol. 4, ed. S. N. Klaus, pp. 291–8, Basel: Karger.

Fanon, F. (1967). *Black Skin, White Masks,* translated by C. L. Markmann. New York: Grove Press.

Farishian, R. A. & Whittaker, J. R. (1980). Phenylalanine lowers melanin synthesis in mammalian melanocytes by reducing tyrosine uptake. *Journal of Investigative Dermatology,* **74**, 85–9.

Farley, R. (1985). Three steps forward and two back? Recent changes in the social and economic status of blacks. *Ethnic and Racial Studies,* **8**, 4–28.

Feeney, L., Grieshaber, J. A. & Hogan, M. J. (1965). Studies on human ocular pigment. In *Eye Structure II,* Symposium, ed. J. W. Rohen, pp. 535–48. Stuttgart: Schattauer-Verlag.

Fellman, J. H. (1958). Epinephrine metabolites and pigmentation in the central nervous system in a case of phenylpyruvic oligophrenia. *Journal of Neurology, Neurosurgery and Psychiatry,* **21**, 58–62.

Fenichel, G. M. & Bazelon, M. (1968). Studies on neuromelanin II: melanin in the brainstems of infants and children. *Neurology,* **18**, 817–20.

Finch, V. A., Dmiel, R., Boxman, R., Shkolnik, A. & Taylor, C. R. (1980). Why black goats in hot deserts? Effects of coat colour on heat exchanges of wild and domestic goats. *Physiological Zoology,* **53**, 19–25.

Findlay, G. H., Morrison, J. G. & Simson, I. W. (1975). Exogenous ochronosis and pigmented colloid milium from hydroquinone bleaching creams. *British Journal of Dermatology,* **93**, 613–22.

Fitzpatrick, T. B. & Breathnach, A. S. (1963). Das epidermale Melanin Einheit-System. *Dermatologische Wochenschrift,* **147**, 481–9.

Fitzpatrick, T. B., Becker, S. W. Jr, Lerner, A. B. & Montgomery, H. (1950). Tyrosinase in human skin: demonstration of its presence and of its role in human melanin formation. *Science,* **112**, 223–5.

Foley, J. H. & Baxter, D. (1958). On the nature of pigment granules in the cells of the locus caeruleus and substantia nigra. *Journal of Neuropathology and Experimental Neurology,* **17**, 586–98.

Follis, R. H., Park, E. A. & Jackson, D. (1952). The prevalence of rickets at autopsy during the first two years of age. *Bulletin of the Johns Hopkins Hospital,* **91**, 480–90.

Ford, J. A., McIntosh, W. B., Butterfield, R. *et al.* (1976). Clinical and subclinical vitamin D deficiency in Bradford children. *Archives of Diseases of Childhood,* **51**, 939–43.

Fox, H. M. & Vevers, G. (1960). *The Nature of Animal Colours.* London: Sidgwick and Jackson.

Fredrickson, G. M. (1981). *White Supremacy: A Comparative Study in American and South African History.* New York: Oxford University Press.

Freedman, B. J. (1984). Caucasian. *British Medical Journal,* **288**, 696–8.

Friedmann, P. S. & Thody, A. J. (1986). Disorders of pigmentation. In *Scientific Basis of Dermatology,* ed. A. J. Thody & P. S. Friedmann, pp. 244–61. Edinburgh: Churchill Livingstone.

Frisancho, A. R., Wainwright, R. & Way, A. (1981). Heritability components of phenotypic expression in skin reflectance of Mestizos from the Peruvian lowlands. *American Journal of Physical Anthropology,* **55**, 203–8.

Fry, L. & Almeyda, J. R. (1968). The incidence of buccal pigmentation in Caucasoids and Negroids in Britain. *British Journal of Dermatology,* **80**, 244–7.

Fuller, R. L. & Geis, S. (1985). The significance of skin color of a newborn infant. *American Journal of Diseases of Children,* **139**, 672–3.

Garber, S. R., Turner, C. W., Creel, D. & Witkop, C. J. (1982). Auditory system abnormalities in human albinos. *Ear and Hearing,* **3**, 207–10.

Garcia, R. I., Flynn, E. & Szabo, G. (1979). Ultrastructure of melanocyte-keratinocyte interactions. In *Pigment Cell,* Vol. 4, ed. S. N. Klaus, pp. 299–307. Basel: Karger.

Garcia, R. I., Mitchell, R. E., Bloom, J. & Szabo, G. (1977). Number of epidermal melanocytes, hair follicles, and sweat ducts in skin of Solomon Islanders. *American Journal of Physical Anthropology,* **47**, 427–34.

   (1983). Solomon Islander skin pigmentation: ultrastructural differences related to genetic variation in Melanesia. *American Journal of Physical Anthropology,* **60**, 323–6.

Garn, S. N. & French, N. Y. (1963). Post-partum and age changes in areolar pigmentation. *American Journal of Obstetrics and Gynaecology,* **85**, 873–5.

Garrard, G., Harrison, G. A. & Owen, J. J. (1967). Comparative spectrophotometry of skin colour with EEL and Photovolt instruments. *American Journal of Physical Anthropology,* **27**, 389–96.

Garrod, A. E. (1908). The Croonian Lectures on inborn errors of metabolism. Lecture 1. *Lancet,* **2**, 1–7.

Gates, R. R. (1938). Blue eyes in natives of Ceylon. *British Medical Journal,* **1**, 921–2.

Gergen, K. J. (1968). The significance of skin color in human relations. In *Color and Race,* ed. J. H. Franklin, pp. 112–28. Boston: Beacon Press.

Gibson, I. M. (1971). Measurement of skin colour *in vivo. Journal of the Society of Cosmetic Chemists,* **22**, 725–40.

Gilchrest, B. A., Blog, F. B. & Szabo, G. (1979). Effects of aging and chronic sun exposure on melanocytes in human skin. *Journal of Investigative Dermatology,* **73**, 141–3.

Glimcher, M. E., Kostick, R. M. & Szabo, G. (1973). The epidermal melanin system in newborn human skin. *Journal of Investigative Dermatology,* **61**, 344–7.

Goel, K. M., Sweet, E. M., Logan, R. W. *et al.* (1976). Florid and subclinical rickets among immigrant children in Glasgow. *Lancet,* **1**, 1141–8.

Goldgeiger, M. H., Klein, L. E., Klein-Angerer, S. *et al.* (1984). The distribution of melanocytes in the leptomeninges of the human brain. *Journal of Investigative Dermatology,* **82**, 235–8.

Goldschmidt, H. & Raymond, J. Z. (1972). Quantitative analysis of skin color from melanin content of superficial skin cells. *Journal of Forensic Sciences,* **17**, 124–31.

Goolamali, S. K. (1973). Levodopa in vitiligo. *Lancet,* **1**, 675–6.

Gordon, P. R. & Gilchrest, B. A. (1989). Human melanogenesis is stimulated by diacylglycerol. *Journal of Investigative Dermatology,* **93**, 700–2.

Gordon, P. R., Mansur, C. P. & Gilchrest, B. A. (1989). Regulation of human melanocyte growth, dendricity, and melanization by keratinocyte derived factors. *Journal of Investigative Dermatology,* **92**, 565–72.

Gould, S. J. (1981). *The Mismeasure of Man.* New York: Norton.

Grainger, K. M. (1973). Pigmentation in Parkinson's disease treated with levodopa. *Lancet,* **1**, 97–8.

Greiner, A. C. & Berry, K. (1964). Skin pigmentation and corneal and lens opacities with prolonged chlorpromazine therapy. *Canadian Medical Association Journal,* **90**, 663–5.

Greiner, A. C. & Nicolson, G. A. (1964). Pigment deposition in viscera associated with prolonged chlorpromazine therapy. *Canadian Medical Association Journal,* **91**, 627–35.

Greiner, A. C. & Nicolson, G. A. (1965). Schizophrenia-melanosis: cause or side-effect? *Lancet,* **2**, 1165–7.

Gureje, O. (1987). Tardive dyskinesia in schizophrenics: prevalence, distribution and relationship to neurological 'soft' signs in Nigerian patients. *Acta Psychiatrica Scandinavica,* **76**, 523–8.

Guthrie, R. D. (1970). Evolution of human threat display organs. In *Evolutionary Biology,* Vol. 4, ed. T. Dobzhansky, M. K. Hecht & W. C. Steere, pp. 257–302. New York: Appleton-Century-Crofts.

Hale, B. D., Landers, D. M., Snyder Bauer, R. & Goggin, N. L. (1980). Iris pigmentation and fractionated reaction and reflex time. *Biological Psychology,* **10**, 57–67.

Halprin, K. M. & Ohkawara, A. (1966). Glutathione and pigmentation. *Archives of Dermatology,* **94**, 355–7.

Hamilton, W. J. & Heppner, F. (1967). Radiant solar energy and the function of black homeotherm pigmentation: an hypothesis. *Science,* **155**, 196–7.

Hamilton, W. J. (1973). *Life's Color Code,* pp. 204–6. New York: McGraw-Hill.

Hanna, J. M. & Baker, P. T. (1974). Comparative heat tolerance of Shipibo Indians and Peruvian Mestizos. *Human Biology,* **46**, 69–80.

Hanna, J. M. & Smith, R. M. (1975). Responses of Hawaiian-born Japanese and Caucasians to a standardized cold exposure. *Human Biology,* **47**, 427–40.

Harburg, E., Erfurt, J. C., Hauenstein, L. *et al.* (1973). Socioecological stress, suppressed hostility, skin color and black–white male blood pressure. *Psychosomatic Medicine,* **35**, 276–96.

Harburg, E., Gleibermann, L., Roeper, P. *et al.* (1978a). Skin color, ethnicity, and blood pressure I. Detroit blacks. *American Journal of Public Health,* **68**, 1177–83.

Harburg, E., Gleibermann, L., Ozgoren, F. *et al.* (1978b). Skin color, ethnicity, and blood pressure II. Detroit whites. *American Journal of Public Health*, **68**, 1184–8.

Harmse, N. S. (1964). Reflectometry of the bloodless living human skin. *Proceedings of the Koninklijke Nederlandse Akademie van Wetenschappen*, Series C, **67**, 138–43.

Harrison, G. A. (1973). Differences in human pigmentation: measurement, geographic variation, and causes. *Journal of Investigative Dermatology*, **60**, 418–26.

Harrison, G. A. & Owen, J. J. T. (1964). Studies on the inheritance of human skin colour. *Annals of Human Genetics*, **28**, 27–37.

Harrison, G. & Salzano, F. M. (1966). The skin colour of the Caingang and Guarini Indians of Brazil. *Human Biology*, **38**, 104–11.

Harrison, G. A., Kuchemann, C. F., Moore, M. A. *et al.* (1969). The effects of altitudinal variation in Ethiopian populations. *Philosophical Transactions of the Royal Society of London, Series B* (Biological Sciences), **256**, 13–182.

Harrison, G. A., Owen, J. J. T., da Rocha, F. J. & Salzano, F. M. (1967). Skin colour in Southern Brazilian populations. *Human Biology*, **39**, 21–31.

Hart, C. W. & Naunton, R. F. (1964). The ototoxicity of chloroquine phosphate. *Archives of Otolaryngology*, **80**, 407–12.

Harvey, R. G. & Lord, J. M. (1978). Skin colour of the Ainu of Hidaka, Hokkaido, Northern Japan. *Annals of Human Biology*, **5**, 459–67.

Harvey, R. G. (1985). Ecological factors in skin color variation among Papua New Guineans. *American Journal of Physical Anthropology*, **66**, 407–16.

Hearing, V. J. & Jiménez, M. (1987). Mammalian tyrosinase – the critical regulatory control point in melanocyte pigmentation. *International Journal of Biochemistry*, **19**, 1141–7.

Hearing, V. J., Korner, A. M. & Pawelek, J. M. (1982). New regulators of melanogenesis are associated with purified tyrosinase isozymes. *Journal of Investigative Dermatology*, **79**, 16–18.

Hedin, C. A. & Larsson, A. (1978). Physiology and pathology of melanin pigmentation with special reference to the oral mucosa. *Swedish Dental Journal*, **2**, 113–29.

Helm, F. & Milgrom, H. (1970). Can scalp hair suddenly turn white? *Archives of Dermatology*, **102**, 102–3.

Hershkowitz, M. (1977). Penile frostbite, an unforeseen hazard of jogging. *New England Journal of Medicine*, **296**, 178.

Hess, A. F. (1922). Newer aspects of the rickets problem. *Journal of the American Medical Association*, **78**, 1177–83.

(1930). *Rickets including Osteomalacia and Tetany*. London: Henry Kimpton.

Hiernaux, J. (1972). La reflectance de la peau dans une communaute de Sara Madjingay (Republique du Tchad). *L'Anthropologie* (Paris), **76**, 279–300.

Hiernaux, J. & Froment, A. (1976). The correlations between anthropobiological and climatic variables in sub-Saharan Africa: revised estimates. *Human Biology*, **48**, 757–67.

Hiernaux, J., Rudan, P. & Brambati, A. (1975). Climate and the weight–height relationship in sub-Saharan Africa. *Annals of Human Biology*, **2**, 3–12.

Hilding, D. A. & Ginzberg, R. D. (1977). Pigmentation of the stria vascularis. *Acta Otorhinolaryngologica*, **84**, 24–37.

Hill, K. (1982). Hunting and human evolution. *Journal of Human Evolution*, **11**, 521–44.

Hirobe, T., Flynn, E. & Szabo, G. (1986). Interaction between white and black melanocytes and keratinocytes *in vitro*. (Abstract). *Journal of Investigative Dermatology*, **87**, 389.

Hirsch, E., Graybiel, A. M. & Agid, Y. A. (1988). Melanized dopaminergic neurons are differentially susceptible to degeneration in Parkinson's disease. *Nature*, **334**, 345–8.

Hoffman, J. M. (1975). Retinal pigmentation, visual acuity and brightness levels. *American Journal of Physical Anthropology*, **43**, 417–24.

Holden, T. J. (1987). Tardive dyskinesia in long-term hospitalized Zulu psychiatric patients. *South African Medical Journal*, **71**, 88–90.

Holick, M. F., MacLaughlin, J. A., Clark, M. B. *et al.* (1980). Photosynthesis of previtamin D_3 in human skin and the physiologic consequences. *Science*, **210**, 203–5.

Holick, M. F., MacLaughlin, J. A. & Doppelt, S. H. (1981). Regulation of cutaneous previtamin D_3 photosynthesis in man: skin pigment is not an essential regulator. *Science*, **211**, 590–3.

Holick, M. F., Smith, E. & Pincus, S. (1987). Skin as the site of vitamin D synthesis and target tissue for 1,25-dihydroxyvitamin D_3. *Archives of Dermatology*, **123**, 1677–83.

Hollis, B. W. & Pittard, W. B. (1984). Evaluation of the total fetomaternal vitamin D relationship at term: evidence for racial differences. *Journal of Clinical Endocrinology and Metabolism*, **59**, 652–7.

Holmes, A. M., Enoch, B. A., Taylor, J. L. & Jones, M. E. (1973). Occult rickets and osteomalacia amongst the Asian immigrant population. *Quarterly Journal of Medicine*, **42**, 125–49.

Hönigsmann, H., Schuler, G., Aberer, W., Romani, V. & Wolff, K. (1986). Immediate pigment darkening phenomenon: a reevaluation of its mechanism. *Journal of Investigative Dermatology*, **87**, 648–52.

Howard, R. O., McDonald, C. J., Dunn, B. & Creasey, W. A. (1969). Experimental chlorpromazine cataracts. *Investigative Ophthalmology*, **8**, 413–42.

Huizinga, J. (1968). Human biological observations on some African populations of the thorn Savanna belt: I and II. *Proceedings of the Koninklijke Nederlandse Akademie van Wetenschappen*, Series C, **71**, 356–90.

Hulse, F. S. (1967). Selection for skin color among the Japanese. *American Journal of Physical Anthropology*, **27**, 143–56.

(1969). Skin color among the Yemenite Jews of the isolate from Habban. *Proceedings of 8th Congress of Anthropological and Ethnological Sciences* (Tokyo), pp. 226–8.

(1973). Skin colour in Northumberland. In *Genetic Variation in Britain*; Symposia of the Society for the Study of Human Biology, Vol. 12, ed. D. F. Roberts & E. Sunderland, pp. 245–57. London: Taylor & Francis.

Hutchinson, J. C. & Brown, G. D. (1969). Penetrance of cattle coats by radiation. *Journal of Applied Physiology*, **26**, 454–64.

Irons, E. (1971). The white Negroes. *Sunday Times Magazine* (London, 17 January), pp. 18–22.

Isaacs, R. H. (1963). Blackness and whiteness. *Encounter, 21*, 8–21.

Ivanhoe, F. (1970). Was Virchow right about Neanderthal? *Nature, 227*, 577–9.

Iwata, M., Corn, T., Iwata, S., Everett, M. A. & Fuller, B. B. (1990). The relationship between tyrosinase activity and skin color in human foreskins. *Journal of Investigative Dermatology, 95*, 9–15.

Jacob, A. I., Sallman, A., Santiz, Z. & Hollis, B. W. (1984). Defective photoproduction of cholecalciferol in normal and uremic humans. *Journal of Nutrition, 114*, 1313–19.

Jaswal, I. J. (1979). Skin colour in northern Indian populations. *Journal of Human Evolution, 8*, 361–6.

Jelinek, J. E. (1972). Sudden whitening of the hair. *Bulletin of the New York Academy of Medicine, 48*, 1003–13.

Jenkins, T., Zoutendyk, A. & Steinberg, A. G. (1970). Gammaglobulin groups (Gm and Inv) of various Southern African populations. *American Journal of Physical Anthropology, 32*, 197–218.

Jimbow, K., Fitzpatrick, T. B., Szabo, G. & Hori, Y. (1975). Congenital circumscribed hypomelanosis. *Journal of Investigative Dermatology, 64*, 50–62.

Jimbow, K., Ishida, O., Ito, S. *et al.* (1983). Combined chemical and electron microscopic studies of pheomelanosomes in human red hair. *Journal of Investigative Dermatology, 81*, 506–11.

Jimbow, K., Oikawa, O., Sugiyama, S. & Takeuchi, T. (1979). Comparison of eumelanogenesis and pheomelanogenesis in retinal and follicular melanocytes: role of vesiculo-globular bodies in melanosome differentiation. *Journal of Investigative Dermatology, 73*, 278–84.

Jimbow, K., Pathak, M. A., Szabo, G. & Fitzpatrick, T. B. (1974). Ultrastructural changes in human melanocytes after ultraviolet radiation. In *Sunlight and Man,* ed. T. B. Fitzpatrick *et al.,* pp. 195–215. Tokyo: University of Tokyo Press.

Jimbow, K., Quevedo, W. C., Fitzpatrick, T. B. & Szabo, G. (1976). Some aspects of melanin biology: 1950–1975. *Journal of Investigative Dermatology, 67*, 72–89.

Johnson, F. S., Mo, T. & Green, A. E. (1976). Average latitudinal variation in ultraviolet radiation at the earth's surface. *Photochemistry and Photobiology, 23*, 179–88.

Johnston, R. F. & Selander, R. K. (1964). House sparrows: rapid evolution of races in North America. *Science, 144*, 548–50.

Jordan, W. D. (1968). *White Over Black: American Attitudes Towards the Negro, 1550–1812.* Chapel Hill: University of North Carolina Press.

Joseph, R. & Godson, P. (1988). Peace at last for tragic Rita: the white outcast trapped in a black skin. *Sunday Times,* Johannesburg, 28 August, p. 12.

Kahlon, D. P. (1973). Skin colour in the Sikh community in Britain. In *Genetic Variation in Britain,* Symposia of the Society for the Study of Human Biology, Vol. 12, ed. D. F. Roberts & E. Sunderland, pp. 265–76. London: Taylor & Francis.

Kaidbey, K. H., Agin, P. P., Sayre, R. M. & Kligman, A. M. (1979). Photoprotection by melanin – a comparison of black and Caucasian skin. *Journal of the American Academy of Dermatology*, **1**, 249–60.

Kaidbey, K. H. (1990). The photoprotective potential of the new superpotent sunscreens. *Journal of the American Academy of Dermatology*, **22**, 449–52.

Kalla, A. K. (1968). Inheritance of skin colour in man. *Anthropologist*, Special Volume: 159–68.

(1969). Affinities in skin pigmentation of some Indian populations. *Human Heredity*, **19**, 499–505.

(1972). Parent–child relationship and sex differences in skin tanning potential in man. *Human genetik*, **15**, 39–43.

(1973). Ageing and sex differences in human skin pigmentation. *Zeitschrift für Morphologie und Anthropologie*, **65**, 29–33.

Kalla, A. K. & Tiwari, S. C. (1970). Sex differences in skin colour in man. *Acta Geneticae Medicae et Gemellologicae*, **19**, 472–6.

(1972). Parent–child relationship and sex differences in skin tanning potential in man. *Humangenetik*, **15**, 39–43.

Kastin, A. J., Kuzemchak, B., Tompkins, R. G. *et al.* (1976). Melanin in rat brain. *Brain Research Bulletin*, **1**, 567–72.

Keeler, C. E. (1963). Caribe Cuna moon-child albinos. *National Cancer Institute Monographs*, **10**, 197–214.

Keil, J. E., Sandifer, S. H., Loadholt, C. B. & Boyle, E. (1981). Skin color and education effects on blood pressure. *American Journal of Public Health*, **71**, 532–4.

Kessler, I. I. (1978). Parkinson's disease in epidemiologic perspective. In *Advances in Neurology*, Vol. 19, ed. B. S. Schoenberg, pp. 355–84. New York: Raven Press.

Kettlewell, B. (1973). *The Evolution of Melanism*, p. 6. Oxford: Clarendon Press.

Khilnani, G., Swaroop, A. K., Goyal, R. K. & Mathur, R. N. (1979). Extrapyramidal reactions due to chloroquin and phenothiazines. *Journal of the Association of Physicians of India*, **27**, 731–4.

King, R. A., Creel, D., Cervenka, J. *et al.* (1980). Albinism in Nigeria with delineation of new recessive oculocutaneous type. *Clinical Genetics*, **17**, 259–70.

King, R. A. & Olds, D. P. (1985). Hairbulb tyrosinase activity in oculocutaneous albinism. *American Journal of Medical Genetics*, **20**, 49–55.

Kiple, K. F. & King, V. H. (1981). *Another Dimension to the Black Diaspora: Diet, Disease and Racism*. Cambridge: Cambridge University Press.

Klaus, S. N. (1969). Pigment transfer in mammalian epidermis. *Archives of Dermatology*, **100**, 756–62.

Kligman, L. H., Akin, F. J. & Kligman, A. M. (1985). The contributions of UVA and UVB to connective tissue damage in hairless mice. *Journal of Investigative Dermatology*, **84**, 272–6.

Knip, A. S. (1977). Ethnic studies on sweat gland counts. In *Physiological Variation and its Genetic Basis*, ed. J. S. Weiner, pp. 113–23 (Symposia of the Society for the Study of Human Biology). London: Taylor and Francis.

Konrad, K. & Wolff, K. (1973). Hyperpigmentation, melanosome size and distribution patterns of melanosomes. *Archives of Dermatology,* **107**, 853–60.

Korner, A. & Pawelek, J. (1982). Mammalian tyrosinase catalyses three reactions in the biosynthesis of melanin. *Science,* **217**, 1163–5.

Kromberg, J. G. (1985). A Genetic and Psychosocial Study of Albinism in Southern Africa. Unpublished Ph.D. thesis, University of the Witwatersrand, Johannesburg.

Kromberg, J. G. & Jenkins, T. (1982). Prevalence of albinism in the South African Negro. *South African Medical Journal,* **61**, 383–6.

(1984). Albinism in the South African Negro III: genetic counselling issues. *Journal of Biosocial Science,* **16**, 99–108.

Kromberg, J. G., Zwane, E. M. & Jenkins, T. (1987). The response of black mothers to the birth of an albino infant. *American Journal of Diseases of Children,* **141**, 911–16.

Kromberg, J. G., Castle, D., Zwane, E. M. & Jenkins, T. (1989). Albinism and skin cancer in South Africa. *Clinical Genetics,* **36**, 43–52.

Kuiters, G. R., Hup, J. M., Siddiqui, A. H. & Cormane, R. H. (1986). Oral phenylalanine loading and sunlight as source of UVA irradiation in vitiligo on the Caribbean island of Curacao NA. *Journal of Tropical Medicine and Hygiene,* **89**, 149–55.

Kung, E. C., Pryson, R. A. & Lenschow, D. H. (1964). Study of a continental surface albedo on the basis of flight measurements and structure of the earth's surface cover over North America. *Monthly Weather Review,* **92**, 543–64.

Kushelevsky, A. P. & Slifkin, M. A. (1975). Ultraviolet measurements at the Dead Sea and at Beersheba. *Israel Journal of Medical Sciences,* **11**, 488–90.

Kuske, H. & Krebs, A. (1964). Hyperpigmentation of chloasma type after treatment with hydantoin preparations. *Dermatologica,* **129**, 121–39.

La Hoste, G. J., Olson, G. A., Kastin, A. J. & Olson, R. D. (1980). Behavioral effects of melanocyte stimulating hormone. *Neuroscience and Biobehavioral Reviews,* **4**, 9–16.

Lacy, M. E. (1984). Phonon–electron coupling as a possible transducing mechanism in bioelectronic processes involving neuromelanin. *Journal of Theoretical Biology,* **111**, 201–4.

Ladell, W. (1964). Terrestrial animals in humid heat. In *Handbook of Physiology: Adaptation to the Environment,* Vol. 4, ed. D. Dill, pp. 625–59. Baltimore: Williams & Wilkins.

Lancet (1980). Hypertension in blacks and whites. Editorial. *Lancet,* **2**, 73–4.

(1981). Sunscreens, photocarcinogenesis, melanogenesis and psoralens. Editorial. *Lancet,* **2**, 335–6.

(1983). Skin photobiology. Editorial. *Lancet,* **1**, 566–8.

(1986). Sunlight and the Granstein cell. Editorial. *Lancet,* **2**, 259–60.

(1987). Sunburn and melanoma. Editorial. *Lancet,* **1**, 1184.

Landers, D. M., Obermeier, G. E. & Patterson, A. H. (1976). Iris pigmentation and reactive motor performance. *Journal of Motor Behavior,* **8**, 171–9.

Lasker, G. W. (1954). Seasonal changes in skin color. *American Journal of Physical Anthropology,* **12**, 553–7.

Lazenby, R. A. & McCormack, P. (1985). Salmon and malnutrition on the northwest coast. *Current Anthropology*, **26**, 379–83.

Lee, M. M. & Lasker, G. W. (1959). The sun-tanning potential of human skin. *Human Biology*, **31**, 252–60.

Lee, R., Mathews-Roth, M. M., Pathak, M. A. & Parrish, J. A. (1975). The detection of carotenoid pigments in human skin. *Journal of Investigative Dermatology*, **64**, 175–7.

Lees, F. C. & Byard, P. J. (1978). Skin colorimetry in Belize. I. Conversion formulae. *American Journal of Physical Anthropology*, **48**, 515–22.

Lees, F. C., Byard, P. J. & Relethford, J. H. (1978). Interobserver error in human skin colorimetry. *American Journal of Physical Anthropology*, **49**, 35–8.

(1979). New conversion formulae for light-skinned populations using photovolt and E.E.L. reflectometers. *American Journal of Physical Anthropology*, **51**, 403–8.

Leguebe, A. (1961). Contribution a l'étude de la pigmentation chez l'homme. *Bulletin de l'Institut royale des sciences naturelles de Belgique*, **37**, 1–29.

Lerner, A. B. (1971). On the etiology of vitiligo and gray hair. *American Journal of Medicine*, **51**, 141–7.

Lerner, A. B. & Lee, T. H. (1955). Isolation of homogenous melanocyte stimulating hormone from hog pituitary glands. *Journal of the American Chemical Society*, **77**, 1066–7.

Lerner, A. B., Case, J. D., Takahasi, Y., Lee, T. H. & Mori, W. (1958). Isolation of melatonin, the pineal gland factor that lightens melanocytes. *Journal of the American Chemical Society*, **80**, 2587.

Lerner, A. B., Halaban, R., Klaus, S. N. & Moellmann, G. E. (1987). Transplantation of human melanocytes. *Journal of Investigative Dermatology*, **89**, 219–24.

Lerner, A. B., Shizume, K. & Bunding, I. (1954). The mechanism of endocrine control of melanin pigmentation. *Journal of Clinical Endocrinology and Metabolism*, **14**, 1463–90.

Lerner, M. R. & Goldman, R. S. (1987). Skin colour, MPTP, and Parkinson's disease. *Lancet*, **2**, 212.

Levantine, A. & Almeyda, J. (1973). Drug induced changes in pigmentation. *British Journal of Dermatology*, **89**, 105–12.

Levy, H. (1982). Chloroquine-induced pigmentation. *South African Medical Journal*, **62**, 735–7.

Lewis, B. (1970). *Race and Color in Islam*. New York: Harper & Row.

Lewis, M. G. (1969). Melanoma and pigmentation of the leptomeninges in Ugandan Africans. *Journal of Clinical Pathology*, **22**, 183–6.

Lieberman, H. R., Waldhauser, F., Garfield, G., Lynch, H. J. & Wurtman, R. J. (1984). Effects of melatonin on human mood and performance. *Brain Research*, **323**, 201–7.

Lin, K-M., Poland, R. E., Nuccio, I. *et al.* (1989). A longitudinal assessment of haloperidol doses and serum concentrations in Asian and Caucasian schizophrenic patients. *American Journal of Psychiatry*, **146**, 1307–11.

Lindquist, N. G. (1973). Accumulation of drugs on melanin. *Acta Radiologica*, supplement **325**, 1–92.

Lindquist, N. G., Larsson, B. S., Lyden-Sokolowski, A. (1987). Neuromelanin and its possible protective and destructive properties. *Pigment Cell Research*, **1**, 133–6.

Little, M. A. & Sprangel, C. J. (1980). Skin reflectance relationships with temperature and skinfolds. *American Journal of Physical Anthropology*, **52**, 145–51.

Little, M. A. & Wolff, M. E. (1981). Skin and hair reflectance in women with red hair. *Annals of Human Biology*, **8**, 231–41.

Livingstone, F. B. (1969). Polygenic models for the evolution of human skin colour differences. *Human Biology*, **41**, 480–93.

Lo, C. W., Paris, P. W. & Holick, M. F. (1986). Indian and Pakistani immigrants have the same capacity as Caucasians to produce vitamin D in response to ultraviolet radiation. *American Journal of Clinical Nutrition*, **44**, 683–5.

Lodge Patch, J. C. (1975). Pigmentation and the chemical basis of schizophrenia. *Acta Psychiatrica Scandinavica*, **51**, 312–18.

Lohr, J. B., Wisniewski, A. & Jeste, D. V. (1986). Neurological aspects of tardive dyskinesia. In *Handbook of Schizophrenia*, Vol. 1, ed. H. A. Nasrallah & D. R. Weinberg, pp. 97–119. Amsterdam: Elsevier.

Loomis, W. F. (1967). Skin-pigment regulation of vitamin D biosynthesis in man. *Science*, **157**, 501–6.

Lorimer, D. A. (1978). *Colour, Class and the Victorians: English attitudes to the Negro in the Mid-Nineteenth Century*. Leicester: Leicester University Press.

Lowenthal, D. (1972). *West Indian Societies*. London: Oxford University Press.

Luande, J., Henschke, C. I. & Mohammed, N. (1985). The Tanzanian human albino skin: natural history. *Cancer*, **55**, 1823–8.

Lustick, S. I. (1969). Bird energetics: effects of artificial radiation. *Science*, **163**, 387–90.

MacCrone, I. D. (1957). *Race Attitudes in South Africa*, p. 293. Johannesburg: Witwatersrand University Press.

MacLaughlin, J. A., Anderson, R. R. & Holick, M. F. (1982). Spectral character of sunlight modulates photosynthesis of previtamin D_3 and its photoisomers in human skin. *Science*, **216**, 1001–3.

Manganyi, N. C., Kromberg, J. G. & Jenkins, T. (1974). Studies on albinism in the South African Negro 1. Intellectual maturity and body image differentiation. *Journal of Biosocial Science*, **6**, 107–12.

Mann, D. M. & Yates, P. O. (1974). Lipoprotein pigments – their relationship to ageing in the human nervous system. *Brain*, **97**, 489–98.

(1979). The effects of ageing on the pigmented nerve cells of the human locus caeruleus and substantia nigra. *Acta Neuropathologica*, **47**, 93–7.

Marsden, C. D. (1961). Pigmentation in the nucleus substantiae nigrae of mammals. *Journal of Anatomy*, **95**, 256–61.

(1965). Brain pigment and catecholamines. *Lancet*, **2**, 1244.

(1983). Neuromelanin and Parkinson's disease. *Journal of Neural Transmission*, supplement **19**, 121–41.

Mason, H. S. (1967). The structure of melanin. In *Advances of Biology of Skin*, Vol. 8, ed. W. Montagna & H. Fu, pp. 293–312. Oxford: Pergamon.

Mason, H. S., Ingram, D. J. & Allen, B. (1960). The free radical property of melanins. *Archives of Biochemistry and Biophysics*, **86**, 225–30.

Mason P. (1968). . . . but O! my soul is white. *Encounter,* **30**, 57–61.

Masson, P. (1948). Pigment cells in man. In *The Biology of Melanomas.* Special publication, New York Academy of Sciences, Vol. 4, 15–37.

Matheny, A. P. & Dolan, A. B. (1975a). Sex and genetic differences in hair color changes during early childhood. *American Journal of Physical Anthropology,* **42**, 53–6.

(1975b). Changes in eye colour during early childhood: sex and genetic differences. *Annals of Human Biology,* **2**, 191–6.

Matsuoko, L. Y., Ide, L., Wortsman, J., MacLaughlin, J. A. & Holick, M. F. (1987). Sunscreens suppress cutaneous vitamin D_3 synthesis. *Journal of Clinical Endocrinology and Metabolism,* **64**, 1165–6.

Matsuoka, L. Y., Wortsman, J., Hanifan, N. & Holick, M. F. (1988). Chronic sunscreen use decreases circulating concentrations of 25-hydroxyvitamin D. *Archives of Dermatology,* **124**, 1802–4.

Mawer, E. B., Backhouse, J., Holman, C. A. *et al.* (1972). The distribution and storage of vitamin D and its metabolites in human tissues. *Clinical Science,* **43**, 413–31.

Mayr, E. & Campbell, B. (1971). Was Virchow right about Neanderthal? *Nature,* **229**, 253–4.

McCord, C. P. & Allen, F. P. (1917). Evidence associating the pineal gland function with alterations of pigmentation. *Journal of Experimental Zoology,* **23**, 207–24.

McGinness, J. (1985). A new view of pigmented neurons. *Journal of Theoretical Biology,* **115**, 475–6.

McGinness, J. E., Corry, P. M. & Proctor, P. H. (1974). Amorphous semiconductor switching in melanins. *Science,* **183**, 853–4.

McGuinness, B. W. (1961). Skin pigmentation and the menstrual cycle. *British Medical Journal,* **2**, 563–5.

Mehrai, H. & Sunderland, E. (1990). Skin colour data from Nowshahr City, Northern Iran. *Annals of Human Biology,* **17**, 115–20.

Menon, I. A. & Haberman, H. F. (1977). Mechanisms of action of melanins. *British Journal of Dermatology,* **97**, 109–12.

Merimee, T. J., Zapf, J. & Froesch, E. R. (1981). Dwarfism in the Pygmy: an isolated deficiency of insulin-like growth factor 1. *New England Journal of Medicine,* **305**, 965–8.

Metchnikoff, E. (1901). On the process of hair turning white. *Proceedings of the Royal Society of Medicine,* **69**, 156.

Milner, D. (1983). *Children and Race: Ten Years On.* London: Ward Lock Educational.

Mitchell, F. T. (1930). Incidence of rickets in the South. *Southern Medical Journal,* **23**, 228.

Mitchell, R. E. (1968). The skin of Australian Aborigines; a light and electron-microscopic study. *Australian Journal of Dermatology,* **9**, 314–28.

Miyamoto, M. & Fitzpatrick, T. B. (1957a). Competitive inhibition of mammalian tyrosinase by phenylalanine and its relationship to hair pigmentation in phenylketonuria. *Nature,* **179**, 199–200.

(1957b). On the nature of pigment in retinal pigment epithelium. *Science,* **126**, 449–50.

Molokhia, M. M. & Portnoy, B. (1973). Trace elements and skin pigmentation. *British Journal of Dermatology*, **89**, 207–9.

Mullen, R. W. (1973). *Blacks in America's Wars*. New York: Monad Press.

Murray, F. G. (1934). Pigmentation, sunlight and nutritional disease. *American Anthropologist*, **36**, 438–45.

Naughton, G. K., Eisinger, M. & Bystryn, J-C. (1983). Detection of antibodies to melanocytes in vitiligo by specific immuno-precipitation. *Journal of Investigative Dermatology*, **81**, 540–2.

Neer, R. M. (1985). Environmental light: effects on vitamin D synthesis and calcium metabolism in humans. *Annals of the New York Academy of Sciences*, **453**, 14–20.

Newman, R. W. (1970). Why man is such a sweaty and thirsty naked animal: a speculative review. *Human Biology*, **42**, 12–27.

Nordlund, J. J. & Lerner, A. B. (1977). The effects of oral melatonin on skin color and on the release of pituitary hormones. *Journal of Clinical Endocrinology and Metabolism*, **45**, 768–74.

Ohmart, R. D. & Lasiewski, R. C. (1971). Roadrunners: energy conservation by hypothermia and absorption of sunlight. *Science*, **172**, 67–9.

Ojikutu, R. O. (1965). Die Rolle von Hautpigment und Schweissdrüsen in der Klimaanpassung des Menschen. *Homo*, **16**, 77–95.

Okazaki, K., Uzuka, M., Morikawa, F., Toda, K. & Seiji, M. (1976). Transfer mechanism of melanosome in epidermal cell culture. *Journal of Investigative Dermatology*, **67**, 541–7.

Okoro, A. N. (1975). Albinism in Nigeria: a clinical and social study. *British Journal of Dermatology*, **92**, 485–92.

Okun, M., Edelstein, L. M., Patel, R. P. & Donnellan, B. (1973). A revised concept of mammalian melanogenesis. *Yale Journal of Biology and Medicine*, **46**, 535–40.

Olson, R. L., Gaylor, J. & Everett, M. A. (1973). Skin color, melanin and erythema. *Archives of Dermatology*, **108**, 541–4.

Oosthuizen, J. M., Theron, J. J., Meyer, A. C. *et al.* (1983). Albinism in blacks – aberrant circadian plasma immunoreactive melatonin levels. *South African Medical Journal*, **64**, 651–2.

Øritsland, N. A. (1971). Wavelength-dependent solar heating of harp seals (*Pagophilus Groenlandicus*). *Comparative Biochemistry and Physiology*, **40A**, 359–61.

Ortner, D. J. (1979). Disease and mortality in the early Bronze Age people of Bab edh-Dhra, Jordan. *American Journal of Physical Anthropology*, **51**, 589–98.

Ortonne, J-P., Mosher, D. B. & Fitzpatrick, T. B. (1983). *Vitiligo and Other Hypomelanoses of Hair and Skin*. New York: Plenum Medical.

Owens, D. W., Knox, J. M., Hudson, H. T. & Troll, D. (1975). Influence of humidity on ultraviolet injury. *Journal of Investigative Dermatology*, **64**, 250–2.

Pakkenberg, H. & Brody, H. (1965). The number of nerve cells in the substantia nigra in paralysis agitans. *Acta Neuropathologica*, **15**, 320–4.

Parkes, J. D., Vollum, D., Marsden, C. D. & Branfoot, A. C. (1972). Café-au-lait spots and vitiligo in Parkinson's disease. *Lancet*, **2**, 1373.

Pathak, M. A. (1982). Sunscreens: topical and systemic approaches for protection of human skin against harmful effects of solar radiation. *Journal of the American Academy of Dermatology,* **7**, 285–312.

Pathak, M. A. & Fitzpatrick, T. B. (1974). The role of natural photoprotective agents in human skin. In *Sunlight and Man,* ed. M. A. Pathak *et al.*, pp. 725–50. Tokyo: University of Tokyo Press.

Pathak, M. A. & Stratton, K. (1968). Free radicals in human skin before and after exposure to light. *Archives of Biochemistry and Biophysics,* **123**, 468–76.

Pawelek, J. M. (1976). Factors regulating growth and pigmentation of melanoma cells. *Journal of Investigative Dermatology,* **66**, 201–9.

Pawson, I. G. & Petrakis, N. L. (1975). Comparison of breast pigmentation among women of different racial groups. *Human Biology,* **47**, 441–50.

Petrie, K., Conaglen, J. V., Thompson, L. & Chamberlain, K. (1989). Effect of melatonin on jet lag after long haul flights. *British Medical Journal,* **298**, 705–7.

Pfeiffer, S. (1984). Paleopathology in an Iroquoian ossuary, with special reference to tuberculosis. *American Journal of Physical Anthropology,* **65**, 181–9.

Phillips, J. I., Isaacson, C. & Carman, H. (1986). Ochronosis in black South Africans who used skin lighteners. *American Journal of Dermatopathology,* **8**, 14–21.

Pollack, M. H. & Manschreck, T. C. (1986). Oculocutaneous albinism and schizophrenia. *Biological Psychiatry,* **21**, 830–3.

Pomerantz, S. H. & Ances, I. G. (1975). Tyrosinase activity in human skin: influence of race and age in newborn. *Journal of Clinical Investigation,* **55**, 1127–31.

Poskitt, E. M., Cole, T. J. & Lawson, D. E. (1979). Diet, sunlight, and 25-hydroxy vitamin D in healthy children and adults. *British Medical Journal,* **1**, 221–3.

Porter, J., Beuf, A. H., Lerner, A. & Nordlund, J. (1987). Response to cosmetic disfigurement: patients with vitiligo. *Cutis,* **39**, 493–4.

Post, P. W., Daniels, F. & Binford, R. T. (1975). Cold injury and the evolution of 'white' skin. *Human Biology,* **39**, 131–43.

Post, P. W., Krauss, A. N., Waldman, S. & Auld, P. A. M. (1976). Skin reflectance of newborn infants from 25 to 44 weeks gestational age. *Human Biology,* **48**, 541–57.

Post, P. W. & Rao, D. C. (1977). Genetic and environmental determinants of skin color. *American Journal of Physical Anthropology,* **47**, 399–402.

Post, R. H. (1962). Population differences in red and green color vision deficiency. *Eugenics Quarterly,* **9**, 131–46.

Potts, A. M. (1962). The concentration of phenothiazines in the eye of experimental animals. *Investigative Ophthalmology,* **1**, 522–30.

(1964). The reaction of uveal pigment *in vitro* with polycyclic compounds. *Investigative Ophthalmology,* **3**, 405–16.

Pouissant, A. F. & Ladner, J. (1968). Black power: a failure for racial integration – within the civil rights movement. *Archives of General Psychiatry,* **18**, 385–91.

Price, J. S., Burton, J. L., Shuster, S. & Wolff, K. (1976). Control of scrotal colour in the vervet monkey. *Journal of Medical Primatology,* **5**, 296–304.

Prota, G. (1988). Some new aspects of eumelanin chemistry. In *Advances in Pigment Cell Research*, ed. J. T. Bagnara, pp. 101–24. New York: Alan R. Liss.

Quevedo, W. C., Fitzpatrick, T. B., Pathak, M. A. & Jimbow, K. (1975). Role of light in human skin color variation. *American Journal of Physical Anthropology*, **43**, 393–408.

Quevedo, W. C., Fitzpatrick, T. B., Szabo, G. & Jimbow, K. (1987). Biology of melanocytes. In *Dermatology in General Medicine*, 3rd edition, ed. T. B. Fitzpatrick *et al.*, pp. 224–51. New York: McGraw-Hill.

Quevedo, W. C., Szabo, G. & Virks, J. (1969). Influence of age and UV on the population of dopa-positive melanocytes. *Journal of Investigative Dermatology*, **52**, 287–90.

Ranadive, N. S., Shirwadkar, S., Persad, S. & Menon, I. A. (1986). Effects of melanin-induced free radicals on the isolated rat peritoneal mast cells. *Journal of Investigative Dermatology*, **86**, 303–7.

Raper, H. S. (1928). Aerobic oxidases. *Physiological Review*, **8**, 245–82.

Rawles, M. E. (1953). Origin of mammalian pigment cell and its role in the pigmentation of hair. In *Pigment Cell Growth*, ed. M. Gordon, pp. 1–15. New York: Academic Press.

Rebato, E. (1987). Skin colour in the Basque population. *Anthropologischer Anzeiger*, **45**, 49–55.

Reinertson, R. P. & Wheatley, V. R. (1959). Studies on the chemical composition of human epidermal lipids. *Journal of Investigative Dermatology*, **32**, 49–59.

Relethford, J. H. & Lees, F. C. (1981). Admixture and skin color in the transplanted Tlaxcaltecan population of Saltillo, Mexico. *American Journal of Physical Anthropology*, **56**, 259–67.

Relethford, J. H., Stern, M. P., Gaskill, S. P. & Hazuda, H. P. (1983). Social class, admixture, and skin color variation in Mexican-Americans and Anglo-Americans living in San Antonio, Texas. *American Journal of Physical Anthropology*, **61**, 97–102.

Resnick, S. (1967). Melasma induced by oral contraceptive drugs. *Journal of the American Medical Association*, **199**, 601–5.

Rickards, D. A. (1965). The therapeutic effect of melatonin in canine melanosis. *Journal of Investigative Dermatology*, **44**, 13–16.

Rife, D. C. (1967). The inheritance of red hair. *Acta Geneticae Medicae et Gemellologiae*, **16**, 342–9.

Riggs, S. K. & Sargent, F. (1964). Physiological regulation in moist heat by young American Negro and white males. *Human Biology*, **36**, 339–53.

Rigters-Aris, C. A. (1973a). Reflectometrie cutanee des Fali (Cameroun). *Proceedings of the Koninklijke Nederlandse Akademie van Wetenschappen, Series C*, **76**, 500–11.

(1973b). A reflectometric study of the skin in Dutch families. *Journal of Human Evolution*, **2**, 123–36.

Rippey, J. J. & Rippey, E. (1984). Epidemiology of malignant melanoma of the skin in South Africa. *South African Medical Journal*, **65**, 595–8.

Roberts, D. F. (1977). Human pigmentation: its geographical and racial distribution and biological significance. *Journal of the Society of Cosmetic Chemists*, **28**, 329–42.

Roberts, D. F. & Kahlon, D. P. (1972). Skin pigmentation and assortative mating in Sikhs. *Journal of Biosocial Science*, **4**, 91–100.

(1976). Environmental correlations of skin colour. *Annals of Human Biology*, **3**, 11–22.

Roberts, D. F., Kromberg, J. G. & Jenkins, T. (1986). Differentiation of heterozygotes in recessive albinism. *Journal of Medical Genetics*, **23**, 323–7.

Robins, A. H. (1972). Skin melanin concentrations in schizophrenia. *British Journal of Psychiatry*, **121**, 613–17.

(1973). Skin melanin content in blue-eyed and brown-eyed subjects. *Human Heredity*, **23**, 13–18.

(1975). Melanosis after prolonged chlorpromazine therapy. *South African Medical Journal*, **49**, 1521–4.

(1979). Melanin pigmentation, women and levodopa. *Archives of Dermatology*, **115**, 817–18.

(1980). Albinism and schizophrenia. *Dermatologica*, **161**, 69–70.

Robinson, S., Dill, D. B., Wilson, J. W. & Nielsen, M. (1941). Adaptations of white men and Negroes to prolonged work in humid heat. *American Journal of Tropical Medicine*, **21**, 261–87.

Rodgers, A. D. & Curzon, G. (1975). Melanin formation by human brain *in vitro*. *Journal of Neurochemistry*, **24**, 1123–9.

Rorsman, H. & Tegner, E. (1988). Biochemical observations in UV-induced pigmentation. *Photodermatology*, **5**, 30–8.

Rosati, G., Granieri, E., Pinna, L. *et al.* (1980). The risk of Parkinson's disease in Mediterranean people. *Neurology*, **30**, 250–5.

Rosdahl, I. & Rorsman, H. (1983). An estimate of the melanocyte mass in humans. *Journal of Investigative Dermatology*, **81**, 278–81.

Rosdahl, I. & Szabo, G. (1976). Ultrastructure of the human melanocyte system in the newborn, with special reference to racial differences. In *Pigment Cell*, Vol. 3, ed. V. Riley, pp. 1–12. Basel: Karger.

Rosenstreich, S. J., Rich, C. & Volwiler, W. (1971). Deposition in and release of vitamin D3 from body fat: evidence for a storage site in the rat. *Journal of Clinical Investigation*, **50**, 679–87.

Rosenthal, N. E., Jacobsen, F. M., Sack, D. A. *et al.* (1988). Atenolol in seasonal affective disorder: a test of the melatonin hypothesis. *American Journal of Psychiatry*, **145**, 52–6.

Rothman, S., Krysa, H. F. & Smiljanic, A. M. (1946). Inhibitory action of human epidermis on melanin formation. *Proceedings of the Society for Experimental Biology and Medicine*, **62**, 208–9.

Sagebiel, R. W. & Odland, G. F. (1972). Ultrastructural identification of melanocytes in early human embryos. In *Pigmentation: Its Genesis and Control*, ed. V. Riley, pp. 43–50. New York: Appleton-Century-Crofts.

Saper, C. B. & Petito, C. K. (1982). Correspondence of melanin-pigmented neurons in human brain with A1–A14 catecholamine cell groups. *Brain*, **105**, 87–101.

Sato, S., Murphy, G. F., Bernhard, J. D., Mihm, M. C. & Fitzpatrick, T. B. (1981). Ultrastructural and X-ray microanalytical observations of minocycline-related hyperpigmentation of the skin. *Journal of Investigative Dermatology,* **77**, 264–71.

Schallreuter, K. U., Hordinsky, M. K. & Wood, J. M. (1987). Thioredoxin reductase: role in free radical reduction in different hypopigmentation disorders. *Archives of Dermatology,* **123**, 615–19.

Schoenberg, B. S. (1986). Descriptive epidemiology of Parkinson's disease: disease distribution and hypothesis formulation. In *Advances in Neurology,* ed. M. D. Yahr & K. J. Bergmann, pp. 277–83. New York: Raven Press.

Schreiner, R. L., Hannemann, R. E., De Witt, D. P. & Moorehead, H. C. (1979). Relationship of skin reflectance and serum bilirubin: full-term Caucasian infants. *Human Biology,* **51**, 31–40.

Schrott, A. & Spoendlin, H. (1987). Pigment anomaly-associated inner ear deafness. *Acta Otolaryngology,* **103**, 451–7.

Schumacher, P. (1973). The girl who turned brown. *Sunday Times Colour Magazine* (Johannesburg, February 25), pp. 6–10.

Seedat, Y. K., Hackland, D. B. & Mpontshane, J. (1981). The prevalence of hypertension in rural Zulus. *South African Medical Journal,* **60**, 7–10.

Seedat, Y. K. & Seedat, M. A. (1982). An inter-racial study of the prevalence of hypertension in an urban South African population. *Transactions of the Royal Society of Tropical Medicine and Hygiene,* **76**, 62–71.

Seedat, Y. K., Seedat, M. A. & Hackland, D. B. (1982). Biosocial factors and hypertension in urban and rural Zulus. *South African Medical Journal,* **61**, 999–1002.

Seiji, M., Fitzpatrick, T. B. & Birbeck, M. S. C. (1961). The melanosome: a distinctive subcellular particle of mammalian melanocytes and the site of melanogenesis. *Journal of Investigative Dermatology,* **36**, 243–52.

Seiji, M., Yoshida, T., Itakura, H. & Irimajiri, T. (1969). Inhibition of melanin formation by sulfhydryl compounds. *Journal of Investigative Dermatology,* **52**, 280–6.

Sellars, S. & Beighton, P. (1983). The Waardenburg syndrome in deaf children in Southern Africa. *South African Medical Journal,* **63**, 725–8.

Sever, R. J., Cope, F. W. & Polis, B. D. (1962). Generation by visible light of labile free radicals in the melanin granules of the eye. *Science,* **137**, 128–9.

Short, G. B. (1975). Iris pigmentation and photopic visual acuity: a preliminary study. *American Journal of Physical Anthropology,* **43**, 425–33.

Shuster, S., Chadwick, L., Afacan, A. S. & Robertson, M. D. (1981). Serum 25-hydroxy vitamin D in surface and underground coalminers. *British Medical Journal,* **283**, 106.

Silman, R. E., Leone, R. M., Hooper, R. J. & Preece, M. A. (1979). Melatonin, the pineal gland and human puberty. *Nature,* **282**, 301–3.

Singhi, S., Singhi, P. & Singh, M. (1979). Extrapyramidal syndrome following chloroquine therapy. *Indian Journal of Paediatrics,* **46**, 58–60.

Smith, A. G., Goolamali, S. K., Thody, A. J. *et al.* (1977a). Phenothiazine therapy and plasma immunoreactive beta-MSH in schizophrenia and pruritic dermatoses. *British Journal of Dermatology,* **96**, 537–9.

Smith, A. G., Shuster, S., Thody, A. J. & Peberdy, M. (1977b). Chloasma and plasma immunoreactive beta-melanocyte-stimulating hormone. *Journal of Investigative Dermatology,* **68**, 169–70.

Smith, D. J. (1977). *Racial Disadvantage in Britain.* Harmondsworth: Penguin Books.

Smith, J. & Mitchell, R. J. (1973). Skin colour studies in South Wales, the Isle of Man and Cumbria. In *Genetic Variation in Britain,* Symposia of the Society for the Study of Human Biology, Vol. 12, ed. D. F. Roberts & E. Sunderland, pp. 259–64. London: Taylor & Francis.

Smith, P. E. (1916). Experimental ablation of the hypophysis in the frog embryo. *Science,* **44**, 280–2.

Snell, R. S. (1965). Effect of melatonin on mammalian epidermal melanocytes. *Journal of Investigative Dermatology,* **44**, 273–5.

Snell, R. S. & Bischitz, P. G. (1960). The effect of large doses of estrogen and estrogen and progesterone on melanin pigmentation. *Journal of Investigative Dermatology,* **35**, 73–82.

(1963). The melanocyte and melanin in human abdominal wall skin. *Journal of Anatomy,* **97**, 361–76.

Snell, R. S. & Turner, R. (1966). Skin pigmentation in relation to the menstrual cycle. *Journal of Investigative Dermatology,* **47**, 147–55.

Snowden, F. M. (1970). *Blacks in Antiquity: Ethiopians in the Greco-Roman Experience.* Cambridge: Harvard University Press.

Snyder, S. H. & D'Amato, R. J. (1986). MPTP: a neurotoxin relevant to the pathophysiology of Parkinson's disease. *Neurology,* **36**, 250–8.

Specker, B. L., Tsang, R. C. & Hollis, B. W. (1985a). Effect of race and diet on human-milk vitamin D and 25-hydroxyvitamin D. *American Journal of Diseases of Childhood,* **139**, 1134–7.

Specker, B. L., Valanis, B., Hertzberg, V. *et al.* (1985b). Sunshine exposure and serum 25 hydroxyvitamin D concentrations in exclusively breast-fed infants. *Journal of Pediatrics,* **107**, 372–6.

Spritz, R. A., Strunk, K. M., Giebel, L. B. & King, R. A. (1990). Detection of mutations in the tyrosinase gene in a patient with type 1a oculocutaneous albinism. *New England Journal of Medicine,* **322**, 1724–8.

Spurgeon, J. H., Meredith, H. V. & Onuoha, G. B. (1984). Skin color comparisons among ethnic groups of college men. *American Journal of Physical Anthropology,* **64**, 413–18.

Sramek, J. J., Sayles, M. A. & Simpson, G. M. (1986). Neuroleptic dosage for Asians: a failure to replicate. *American Journal of Psychiatry,* **143**, 535–6.

Stamp, T. C. (1975). Factors in human vitamin D nutrition and in the production and cure of classical rickets. *Proceedings of the Nutrition Society,* **34**, 119–30.

Staricco, R. J. & Pincus, H. (1957). Quantitative and qualitative data on the pigment cells of adult human epidermis. *Journal of Investigative Dermatology,* **28**, 33–45.

Steegmann, A. T. (1967). Frostbite of the human face as a selective force. *Human Biology,* **39**, 131–43.

Steegmann, A. T. (1975). Human adaptation to cold. In *Physiological Anthropology,* ed. A. Damon, pp. 130–66. New York: Oxford University Press.

Steel, K. P. & Barkway, C. (1989). Another role for melanocytes: their importance for normal stria vascularis development in the mammalian inner ear. *Development*, **107**, 453–63.

Stenson, S. M., Siegel, I. M. & Carr, R. E. (1983). Infantile cystinosis: ocular findings and pigment dilution of eye and skin. *Ophthalmic Paediatrics and Genetics*, **3**, 169–80.

Stepan, N. (1982). *The Idea of Race in Science*. London: Macmillan.

Stern, C. (1970). Model estimates of the number of gene pairs involved in pigmentation variability of the Negro-American. *Human Heredity*, **20**, 165–8.

Stewart, H. T. & Keeler, C. E. (1965). A comparison of the intelligence and personality of moon-child albino and control Cuna Indians. *Journal of Genetic Psychology*, **106**, 319–24.

Stolar, R. (1963). Induced alterations of vitiliginous skin. *Annals of the New York Academy of Sciences*, **100**, 58–75.

Stierner, U., Rosdahl, I., Augustsson, A. & Kagedal, B. (1988). Urinary excretion of 5-S-cysteinyldopa in relation to skin type, UVB-induced erythema, and melanocyte proliferation in human skin. *Journal of Investigative Dermatology*, **91**, 506–10.

Strydom, N. B. & Wyndham, C. H. (1963). Natural state of heat acclimatization of different ethnic groups. *Federation Proceedings*, **22**, 801–9.

Sunderland, E. (1967). The skin colour of the people of Azraq, Eastern Jordan. *Human Biology*, **39**, 65–70.

Sunderland, E., Tills, D., Bouloux, C. & Doyl, J. (1973). Genetic studies in Ireland. In *Genetic Variation in Britain*, Symposia of the Society for the Study of Human Biology, Vol. 12, ed. D. F. Roberts & E. Sunderland, pp. 141–59. London: Taylor & Francis.

Sunderland, E. & Woolley, V. (1982). A study of skin pigmentation in the population of the former county of Pembrokeshire, Wales. *Human Biology*, **54**, 387–401.

Szabo, G. (1954). The number of melanocytes in human epidermis. *British Medical Journal*, **1**, 1016–17.

Szabo, G., Gerald, A. B., Pathak, M. A. & Fitzpatrick, T. B. (1969). Racial differences in the fate of melanosomes in human epidermis. *Nature*, **222**, 1081–2.

Tasa, G. L., Murray, C. J. & Boughton, J. M. (1985). Reflectometer reports on human pigmentation. *Current Anthropology*, **26**, 511–12.

Tauber, H. (1981). [13]C evidence for dietary habits of prehistoric man in Denmark. *Nature*, **292**, 332–3.

Taylor, W. O. (1980). Albino Fellowship: a new kind of welfare organization. *Practitioner*, **224**, 1184–7.

(1987). Prenatal diagnosis of albinism. *Lancet*, **1**, 1307–8.

Tedford, W. H., Hill, W. R. & Hensley, L. (1978). Human eye colour and reaction time. *Perceptual and Motor Skills*, **47**, 503–6.

Teplin, L. A. (1976). A comparison of racial/ethnic preferences among Anglo, Black and Latino children. *American Journal of Orthopsychiatry*, **46**, 702–9.

Thody, A. J. & Smith, A. G. (1977). Hormones and skin pigmentation in the mammal. *International Journal of Dermatology*, **16**, 657–64.

Thomson, M. L. (1951). The cause of changes in sweating rate after ultraviolet radiation. *Journal of Physiology,* **112**, 31–42.

(1955). Relative efficiency of pigment and horny layer thickness in protecting the skin of Europeans and Africans against solar ultraviolet radiation. *Journal of Physiology,* **127**, 236–46.

Tiwari, S. C. (1963). Studies of crossing between Indians and Europeans. *Annals of Human Genetics,* **26**, 219–27.

Tobias, P. V. (1961). Studies on skin reflectance in Bushman-European hybrids. *Proceedings, Second International Congress of Human Genetics, Rome,* pp. 461–71.

(1974). The biology of the South African Negro. In *The Bantu-Speaking Peoples of Southern Africa,* ed. W. D. Hammond-Tooke, pp. 3–45. London: Routledge & Kegan Paul.

Toda, K., Pathak, M. A., Parrish, J. A. & Fitzpatrick, T. B. (1972). Alterations of racial differences in melanosome distribution in human epidermis after exposure to ultraviolet light. *Nature (New Biology),* **236**, 143–5.

Tomita, Y., Torinuki, W. & Tagami, H. (1988). Stimulation of human melanocytes by vitamin D3 possibly mediates skin pigmentation after sun exposure. *Journal of Investigative Dermatology,* **90**, 882–4.

Tomita, Y., Takeda, A., Okinaga, S., Tagami, H. & Shibahara, S. (1989). Human oculocutaneous albinism caused by single base insertion in the tyrosinase gene. *Biochemical and Biophysical Research Communications,* **164**, 990–6.

Tsafrir, J. S. (1974). *Light-Eyed Negroes and the Klein-Waardenburg Syndrome.* London: Macmillan.

Turner, V. W. (1966). Colour classification in Ndembu ritual. In *Anthropological Approaches to the Study of Religion,* ed. M. Banton, pp. 47–84. London: Tavistock.

Unterhalter, B. (1975). Changing attitudes to 'passing for white' in an urban Coloured community. *Social Dynamics,* **1**, 53–62.

Urbach, F. (1969). Geographic pathology of skin cancer. In *The Biological Effects of Ultraviolet Radiation,* ed. F. Urbach, pp. 635–50. Oxford: Pergamon.

van den Berghe, P. L. & Frost, P. (1986). Skin color preference, sexual dimorphism and sexual selection: a case of gene culture co-evolution? *Ethnic and Racial Studies,* **9**, 87–113.

Van der Westhuyzen, J. (1986). Biochemical evaluation of black preschool children in the northern Transvaal. *South African Medical Journal,* **70**, 146-8.

Van der Westhuyzen, J., van Tonder, S. V., Gilbertson, I. & Metz, J. (1986). Iron, folate and vitamin B_{12} nutrition and anaemia in black preschool children in the northern Transvaal. *South African Medical Journal,* **70**, 143–6.

Van Rijn-Tournel, J. (1966). Pigmentation de la peau de Belges et d'Africains. *Bulletin de la Société royale Belge d'Anthropologie et de Préhistoire,* **76**, 79–96.

Van Woert, M. H. (1970). Effect of phenothiazines on melanoma tyrosinase activity. *Journal of Pharmacology and Experimental Therapeutics*, **173**, 256–64.

Vernall, D. G. (1963). A study of the density of pigment granules in hair from four races of men. *American Journal of Physical Anthropology*, **21**, 489–96.

Vieregge, P., Kömpf, D. & Fassl, H. (1988). Environmental toxins in Parkinson's disease. *Lancet*, **1**, 362–3.

Vik, T., Try, K. & Stromme, J. H. (1980). The vitamin D status of man at 70° north. *Scandinavian Journal of Clinical and Laboratory Investigation*, **40**, 227–32.

Wagatsuma, H. (1968). The social perception of skin color in Japan. In *Color and Race*, ed. J. H. Franklin, pp. 129–65. Boston: Beacon Press.

Waldhauser, F., Weiszenbacher, G., Frisch, H. *et al.* (1984). Fall in nocturnal serum melatonin during prepuberty and pubescence. *Lancet*, **1**, 362–4.

Walsberg, G. E., Campbell, G. S. & King, J. R. (1978). Animal coat color and radiative heat gain: a re-evaluation. *Journal of Comparative Physiology*, **126**, 211–22.

Walsh, R. J. (1964). Variation in the melanin content of the skin of New Guinea natives at different ages. *Journal of Investigative Dermatology*, **42**, 261–5.

Ward, S. H. & Brown, J. (1972). Self-esteem and racial preference in black children. *American Journal of Orthopsychiatry*, **42**, 644–7.

Wassermann, H. P. (1965a). The circulation of melanin – its clinical and physiological significance. *South African Medical Journal*, **39**, 711–16.

(1965b). Human pigmentation and environmental adaptation. *Archives of Environmental Health*, **11**, 691–4.

(1974). *Ethnic Pigmentation: Historical, Physiological and Clinical Aspects.* Amsterdam: Excerpta Medica.

Wassermann, H. P. & Heyl, T. (1968). Quantitative data on skin pigmentation in South African races. *South African Medical Journal*, **42**, 98–101.

Wästerström, S-A. (1984). Accumulation of drugs on inner ear melanin. *Scandinavian Audiology*, supplement **23**, 1–40.

Weatherhead, B. (1982). The pineal gland and pigmentation. In *The Physiology and Pathophysiology of the Skin*, ed. A. Jarrett, pp. 2165–79. London: Academic Press.

Webb, A. R., De Costa, B. R. & Holick, M. F. (1989). Sunlight regulates the cutaneous production of vitamin D_3 by causing its photodegradation. *Journal of Clinical Endocrinology and Metabolism*, **68**, 882–7.

Webb, A. R., Kline, L. & Holick, M. F. (1988). Influence of season and latitude on the cutaneous synthesis of vitamin D_3: exposure to winter sunlight in Boston and Edmonton will not promote vitamin D_3 synthesis in human skin. *Journal of Clinical Endocrinology and Metabolism*, **67**, 373–8.

Weigand, D. A., Haygood, C. & Gaylor, J. P. (1974). Cell layers and density of Negro and Caucasian stratum corneum. *Journal of Investigative Dermatology*, **62**, 563–8.

Weiner, J. S. (1951). A spectrophotometer for measurement of skin colour. *Man*, **51**, 152–3.

(1969). A preliminary report on the Sandawe of Tanzania. Paper presented to Anthropology Congress, Prague, cited by Rigters-Aris (1973a).

Weiner, J. S., Harrison, G. A., Singer, R., Harris, R. & Jopp, W. (1964). Skin colour in Southern Africa. *Human Biology*, **36**, 294–307.

Weiner, J. S. & Lourie, J. A. (1981). Measurement of skin colour. In *Practical Human Biology*, pp. 141–5. London: Academic Press.

Weiner, J. S., Sebag-Montefiore, N. C. & Peterson, J. N. (1963). A note on the skin colour of Aguarana Indians of Peru. *Human Biology*, **35**, 470–3.

Wells, C. (1975). Prehistoric and historical changes in nutritional diseases and associated conditions. *Progress in Food and Nutrition Science*, **1**, 729–79.

Weninger, M. (1969). Spektrophotometrische Untersuchungen der Haut an einem Bantu-Stamm (Chope) aus Mocambique. *Anthropologie*, **7**, 53–8.

Westerhof, W., Pavel, S., Kammeyer, A. & Beusenberg, F. D. (1987). Melanin-related metabolites as markers of the skin pigmentary system. *Journal of Investigative Dermatology*, **89**, 78–81.

Wheeler, R. H., Bhalerao, V. R. & Gilkes, M. J. (1969). Ocular pigmentation, extrapyramidal symptoms and phenothiazine dosage. *British Journal of Psychiatry*, **115**, 687–90.

White, C. M. (1954). Some Ibo attitudes to skin pigmentation. *Man*, **54**, 147.

Wienker, C. W. (1979). Skin color in a group of black Americans. *Human Biology*, **51**, 1–9.

Williams, J. E., Boswell, D. A. & Best, D. L. (1975). Evaluative responses of preschool children to the colors white and black. *Child Development*, **46**, 501–8.

Williams, J. E. & Carter, D. J. (1967). Connotations of racial concepts and color names in Germany. *Journal of Social Psychology*, **72**, 19–26.

Williams, J. E., Tucker, R. D. & Dunham, F. Y. (1971). Changes in connotations of color names among Negroes and Caucasians. *Journal of Personality and Social Psychology*, **19**, 222–8.

Wirestrand, L-E., Hansson, C., Rosengren, E. & Rorsman, H. (1985). Melanocyte metabolites in the urine of people of different skin colour. *Acta Dermato-Venereologica*, **65**, 345–8.

Witkop, C. J., Niswander, J. D., Bergsma, D. R. *et al.* (1972). Tyrosinase positive oculocutaneous albinism among the Zuni and the Brandywine triracial isolate. *American Journal of Physical Anthropology*, **36**, 397–406.

Witkop, C. J., Quevedo, W. C. & Fitzpatrick, T. B. (1983). Albinism and other disorders of pigment metabolism. In *The Metabolic Basis of Inherited Disease*, ed. J. B. Stanbury, J. B. Wyngaarden, D. S. Frederickson, J. L. Goldstein & M. S. Brown, pp. 301–46. New York: McGraw-Hill.

Wolf, C., Steiner, A. & Hönigsmann, H. (1988). Do oral carotenoids protect human skin against ultraviolet erythema, psoralen phototoxicity, and ultraviolet-induced DNA damage? *Journal of Investigative Dermatology*, **90**, 55–7.

Worthy, M. (1974). *Eye Color, Sex, and Race: Keys to Human and Animal Behaviors*. Anderson, South Carolina: Droke House/Hallux.

Worthy, M. & Markle, A. (1970). Racial differences in reactive versus self-paced sports activities. *Journal of Personality and Social Psychology*, **16**, 439–43.

Young, A. R., Potten, C. S., Chadwick, C. A., Murphy, G. M. & Cohen, A. J. (1988). Inhibition of UV radiation-induced DNA damage by a 5-methoxypsoralen tan in human skin. *Pigment Cell Research*, **1**, 350–4.

Young, L. & Bagley, C. (1979). Identity, self-esteem and evaluation of colour and ethnicity in young children in Jamaica and London. *New Community*, **7**, 154–69.

Zivanovic, S. (1982). *Ancient Diseases: The Elements of Palaeopathology*, pp. 109–14. London: Methuen.

Zvelebil, M. (1986). Postglacial foraging in the forests of Europe. *Scientific American*, **254**, 86–93.

Index

Page numbers in *italics* refer to figures and tables.